00-CTS-17UR

CHARACTERIZING MECHANISMS OF SIMULTANEOUS BIOLOGICAL NUTRIENT REMOVAL DURING WASTEWATER TREATMENT

by:

Peter F. Strom, Ph.D.
Helen X. Littleton, Ph.D., P.E.
Rutgers, The State University of New Jersey

Glen Daigger, Ph.D., P.E.
CH2M HILL

2004

WERF

The Water Environment Research Foundation, a not-for-profit organization, funds and manages water quality research for its subscribers through a diverse public-private partnership between municipal utilities, corporations, academia, industry, and the federal government. WERF subscribers include municipal and regional water and wastewater utilities, industrial corporations, environmental engineering firms, and others that share a commitment to cost-effective water quality solutions. WERF is dedicated to advancing science and technology addressing water quality issues as they impact water resources, the atmosphere, the lands, and quality of life.

For more information, contact:
Water Environment Research Foundation
635 Slaters Lane, Suite 300
Alexandria, VA 22314-1177
Tel: (703) 684-2470
Fax: (703) 299-0742
www.werf.org
werf@werf.org

This report was co-published by the following organizations. For non-subscriber sales information, contact:

IWA Publishing
Alliance House, 12 Caxton Street
London SW1H 0QS, United Kingdom
Tel: +44 (0) 20 7654 5500
Fax: +44 (0) 20 7654 5555
www.iwapublishing.com
publications@iwap.co.uk

Library of Congress Catalog Card Number: 2003115459

IWAP ISBN: 1-84339-705-6

This report was prepared by the organization(s) named below as an account of work sponsored by the Water Environment Research Foundation (WERF). Neither WERF, members of WERF, the organization(s) named below, nor any person acting on their behalf: (a) makes any warranty, express or implied, with respect to the use of any information, apparatus, method, or process disclosed in this report or that such use may not infringe on privately owned rights; or (b) assumes any liabilities with respect to the use of, or for damages resulting from the use of, any information, apparatus, method, or process disclosed in this report.

Rutgers, the State University of New Jersey

This document was reviewed by a panel of independent experts selected by WERF. Mention of trade names or commercial products does not constitute WERF nor EPA endorsement or recommendations for use. Similarly, omission of products or trade names indicates nothing concerning WERF's nor EPA's positions regarding product effectiveness or applicability.

The research on which this report is based was funded, in part, by the United States Environmental Protection Agency through Cooperative Agreement No. CR-827345-01 with the Water Environment Research Foundation (WERF). Unless an EPA logo appears on the cover, this report is a publication of WERF, not EPA. Funds awarded under the Cooperative Agreement cited above were not used for editorial services, reproduction, printing, or distribution.

ACKNOWLEDGMENTS

The authors would like to thank the following organizations for their support of this work: The Water Environment Research Foundation, U.S. Filter/Envirex, Evesham Municipal Utilities Authority, Mt. Laurel Municipal Utilities Authority, Hammonton Township wastewater treatment plant, CH2M HILL, Rutgers—The State University of New Jersey, and Cook College and the New Jersey Agricultural Experiment Station. We also extend special thanks to the following individuals: Peter Petit (Director of Research), Dr. George Smith, Chuck Applegate, and Jim Ehleiter, all of U.S. Filter/Envirex; Rocco Maiellano, Debbie Waller, Joe Rizzuto, Dave Leback, and all the other operators of the Elmwood wastewater treatment plant; Dr. Carlos Filipe, McMaster University; Dr. Robert Cowan and Rong Jin, Rutgers University.

Report Preparation

Principal Investigator:
Peter F. Strom, Ph.D.
Rutgers, The State University of New Jersey

Project Team:
Glen Daigger, Ph.D., P.E.
CH2M HILL

Helen X. Littleton, Ph.D., P.E.
Rutgers, The State University of New Jersey

Project Subcommittee

Water Environment Research Foundation Staff

ABSTRACT AND BENEFITS

Abstract:

Simultaneous biological nutrient removal (SBNR) is the removal of nitrogen and/or phosphorus in excess of that required for biomass synthesis in biological wastewater treatment systems where there are no defined anaerobic and/or anoxic zones. The hypothesis is that one or more of three mechanisms are responsible within individual systems: variations in the bioreactor macroenvironment created by the mixing pattern, gradients within the floc microenvironment, and/or novel microorganism activity. Understanding the mechanisms of SBNR is likely to lead to applying it more efficiently and reliably.

Preliminary work documented SBNR in seven full-scale Orbal™ closed loop bioreactors. A batch assay demonstrated that novel microorganism activity was of little importance in SBNR at the three plants tested. While the floc microenvironment probably plays an important role in nitrogen removal in such plants, it cannot explain phosphorus removal.

A computational fluid dynamics (CFD) model was developed to elucidate the role of the bioreactor macroenvironment in SBNR. This is the first reported application of CFD to activated sludge biological wastewater treatment. Although the software and computational requirements limited model complexity, it still simulated the creation of dissolved oxygen gradients within the system, demonstrating that the anaerobic zones required for SBNR could occur.

Benefits:

- Demonstrates that the bioreactor macroenvironment created by the mixing pattern in a closed loop bioreactor can contribute to simultaneous biological nutrient removal.
- Shows that a new tool, computational fluid dynamics, can be applied to complex biological wastewater treatment systems.
- Illustrates that oxygen input is an important control parameter for aerated systems even when dissolved oxygen concentrations are below detection.
- Indicates that it is possible to remove biological nutrients more reliably and efficiently.

Keywords: activated sludge, mixing, nitrogen, phosphorus, computational fluid dynamics

TABLE OF CONTENTS

LIST OF TABLES

LIST OF FIGURES

LIST OF ACRONYMS

ASM	IWA's Activated Sludge Model; ASM1, ASM2, and ASM2d refer to Models No. 1, 2, and 2d, respectively
BNR	Biological Nutrient Removal
BOD	Biochemical Oxygen Demand
BOD_5	Five-day Biochemical Oxygen Demand
CFD	Computational Fluid Dynamics
CFX	Commercial CFD Software provided by ANSYS (Pittsburgh, PA); CFX4 and CFX4.4 refer to versions of CFX
COD	Chemical Oxygen Demand
DO	Dissolved Oxygen
Het	Heterotroph (usually referring to heterotrophs other than PAOs)
HRT	Hydraulic Residence Time
IWA	International Water Association
MLSS	Mixed Liquor Suspended Solids
MLVSS	Mixed Liquor Volatile Suspended Solids
N_2	Elemental Nitrogen Gas
NH_4^+-N	Ammonium Nitrogen
NO_2^--N	Nitrite-Nitrogen
NO_3^--N	Nitrate-Nitrogen
ORP	Oxidation-Reduction Potential
PAO	Phosphate Accumulating Organism
PHA	Poly-β-hydroxyalkonate
PP	Polyphosphate
RAS	Return Activated Sludge
ROM	Read Only Memory
SBNR	Simultaneous Biological Nutrient Removal
SRT	Solids Retention Time
SSSP	Single Sludge Simulation Program
TIN	Total (fixed) Inorganic Nitrogen (includes ammonium, nitrite, and nitrate, but not elemental nitrogen)
TKN	Total Kjeldahl Nitrogen (includes organic nitrogen and ammonium)
TP	Total Phosphorus
TPO_4^{3-}-P	Total Phosphate-Phosphorus
TSS	Total Suspended Solids
URF	Under Relaxation Factor
U.S. EPA	United States Environmental Protection Agency
VSS	Volatile Suspended Solids
WERF	Water Environment Research Foundation

EXECUTIVE SUMMARY

ES.1 Background

Activated sludge is the most widely used method for secondary treatment of wastewater. It is a suspended growth aerobic process for biodegrading organic matter and converting organic nitrogen to ammonium. It also is widely used to transform ammonium-nitrogen to nitrate-nitrogen via the biological process of nitrification. More recently it has been configured to reliably achieve nutrient removal by providing mixed, non-aerated zones and internal process recycle streams to create the anoxic and/or anaerobic environments needed for biological nitrogen and/or phosphorus removal. Biological nutrient removal (BNR) activated sludge systems using these well defined anoxic and anaerobic zones have evolved since the 1970s to the point where it is a widely applied wastewater treatment technology.

It has also been noted, however, that nutrient removal occurs in systems that do not have formal (intentionally included) anoxic or anaerobic zones. Simultaneous biological nutrient removal (SBNR) is the removal of nitrogen and/or phosphorus in excess of that required for biomass synthesis in biological wastewater treatment systems where defined anaerobic and/or anoxic zones do not exist.

Three mechanisms have been proposed to account for SBNR: 1) the necessary anoxic and/or anaerobic zones are created informally as a result of the mixing pattern in the **bioreactor macroenvironment** produced by the aeration and mixing equipment; 2) anoxic and/or anaerobic zones develop within the **floc microenvironment** due to the internal diffusion limitations; and 3) **novel microorganisms** are active, with nontraditional biochemical pathways for nitrogen or phosphorus transformations. The basic hypothesis of this overall research is that, while these three general mechanisms may be responsible for SBNR, system design and operating parameters will determine the relative importance of each within a specific system.

ES.2 Project Objectives

Understanding SBNR's mechanisms can be expected to lead to applying it more efficiently and effectively. The objective of this overall research project has been to identify the factors affecting the relative contributions of the three proposed mechanisms to SBNR, in order to implement and control SBNR more effectively in full-scale plants. A key to controlling SBNR is to understand how process design and operating parameters affect each of these three mechanisms.

This report 1) provides the results of preliminary field studies at full-scale Orbal™ plants, studies that indicated that the floc microenvironment plays a role in simultaneous nitrogen but not in phosphorus removal, and 2) describes the development and application of batch assays that found the activity of novel microorganisms to be, at most, of minor importance in SBNR at the plants tested. The report then focuses on the development and application of a computational fluid dynamics (CFD) model as a new tool to elucidate the role of the bioreactor macroenvironment in SBNR, particularly for phosphorus removal. This represents the first reported application of CFD to the biological treatment of wastes.

ES.3 Preliminary Work

During the preliminary field study, the researchers routinely observed extensive total nitrogen removal in full-scale staged, closed-loop bioreactors even though there were no intentionally incorporated formal anaerobic or anoxic zones. In some of the facilities, biological phosphorus removal was also observed. SBNR was accomplished in these three-channel systems by restricting the oxygen input of the outer (first) two channels, resulting in low DO concentrations there. The extent of nutrient removal appeared to be controlled largely by the oxygen input to these channels – the greater the degree of oxygen deficiency, within limits, the greater the nutrient removal.

The International Water Association's (IWA's) Activated Sludge Model No. 1 (ASM1) was used to show that the floc microenvironment mechanism could explain the observed nitrogen removal in these systems. However, because it could not produce alternating anaerobic and aerobic exposures for the phosphate-accumulating organisms (PAOs), this mechanism did not explain the phosphorus removal.

The series of batch assays developed, based on the known biochemistry of the novel microorganisms considered, proved a useful screening tool for the activity of autotrophic denitrification under anoxic conditions and heterotrophic nitrification under aerobic conditions in mixed liquor from full-scale bioreactors accomplishing SBNR. In the three systems tested, there was no evidence of substantial activity by novel nitrogen-transforming microorganisms. Further, the balanced populations of ammonia- and nitrite-oxidizing bacteria present, as evidenced by the lack of nitrite accumulation during the aerobic batch assays, suggest that ammonia is being converted to nitrate in the full-scale systems before it is subsequently denitrified (converted to nitrogen gas). One set of batch test results with a very low aeration rate also demonstrated that nitrification and denitrification could occur simultaneously without detected DO.

Anaerobic/aerobic batch assays assessed the activity of PAOs at one plant. As expected, based on current understanding of PAO metabolism, the researchers observed anaerobic phosphate release (following acetate addition) and aerobic phosphate uptake.. The zero order phosphate uptake rate suggested that PAOs constituted ~13% of the process mixed liquor. Coupled with previous observations of process phosphate removal, mixed liquor phosphate content, and the presence of intracellular polyphosphate granules, these results strongly suggest that excess biological phosphorus removal was occurring in these systems. These results also demonstrate that the mixed liquor was capable of phosphate release, even though such release was not observed in the full-scale system. Thus the researchers found no evidence of novel P uptake microorganisms.

ES.4 Research Approach

Computational fluid dynamics (CFD) is the science of determining a mathematical solution to fluid flow in a system through space and time. By means of computer-based simulation, CFD analyses fluid flow, heat transfer and associated phenomena such as chemical reactions.. The technique is very powerful and spans a wide range of industrial and nonindustrial applications. Generally, the water and wastewater industries have been more concerned with the chemical compounds present than with the physical properties of the fluid in which they react. Although CFD has been used to analyze flow distributions in storage, disinfection contact, and settling tanks, and in pumps and mixing/injecting systems, this is the first report of CFD used directly to examine activated sludge biochemical reactions combined with the fluid mixing pattern.

It is generally recognized that spatially varying environments can develop within bioreactors with high mixed liquor recirculation rates, such as in closed-loop bioreactors and basins that are aerated with surface mechanical aerators. But it has often been assumed that flow in closed-loop bioreactors has a laminar pattern. Preliminary velocity measurements and visual observations in a closed-loop bioreactor test tank suggested instead that a spiral type flow coexists with the main recirculating flow. This "second current" is actually a well-known phenomenon in curved pipe or tank environments.

The researchers hypothesized that this second current in closed-loop bioreactors, along with the main mixed liquor recirculation flow, could create alternating anaerobic and aerobic conditions, both vertically and horizontally, which would allow SBNR to occur. Therefore, understanding these mixing patterns could be crucial to providing reliable nutrient removal, and CFD could detect and characterize such zones.

The research described in this report focused on elucidating the role of the bioreactor macroenvironment in SBNR, especially for phosphorus removal. Using the commercial CFD software package CFX4 (ANSYS, Pittsburgh, PA), the researchers developed a computational fluid dynamics model, for a test tank and for the outer channel of a full-scale Orbal™ closed-loop bioreactor. The simplified approach used to create fluid motion in the two tanks consisted of imparting a fraction of the tank's observed momentum (based on measured velocity) uniformly across the location of the disc aerators in a horizontal direction. Predicted velocity profiles were determined and compared against field measurements. The researchers evaluated the potential for DO variations in the full-scale tank by including a source of oxygen at the location of the disc aerators and an oxygen sink distributed throughout the contents of the full-scale bioreactor.

Biochemical reactions based on IWA's Activated Sludge Model No. 2 (ASM2) then were added for the full-scale system CFD model. The chemical species of interest were introduced into the CFD model via source and sink scalar functions in a FORTRAN subroutine to satisfy the Navier-Stokes series of governing equations for conservation of mass and momentum. The simulations were performed using acetate as the sole carbon and energy source, and with no return sludge. Simulations for general heterotrophs and phosphate-accumulating organisms (PAOs) individually as well as for the competition of the two groups of organisms were conducted separately with varied oxygen input. The researchers compared the results from CFD simulations to ASM2 predictions for a completely mixed tank or series of tanks, and in some cases, to an analytical solution and/or field data.

ES.5 Conclusions

The flow patterns predicted for the test and full-scale tanks yielded a good degree of fit with the measured velocities. Small systematic errors were observed, probably due to the simplifications employed, but overall the CFD model captured well the mixing pattern generated within the tanks. This complex flow pattern consisted of a high velocity discharge from the aerator that flowed downward to the bottom of the tank at the first curve, and then upward to the top of the tank on the side opposite the aerator. The resulting flow thus had a spiral form.

Even with the highly simplified form of the biochemical equations consisting only of an oxygen sink distributed throughout the tank, the CFD model predicted zones of relatively higher and lower DO concentrations occurring in the bioreactor. These results suggested that the heterogeneous flow pattern created in the bioreactor macroenvironment could result in the creation of informal anaerobic and anoxic zones that would allow SBNR to occur in such aerated systems.

The more complex CFD model incorporating growth of microbial populations was able to provide useful simulations to qualitatively describe the bioreactor system. This allowed the researchers to test the hypothesis that the bioreactor macroenvironment contributes to SBNR, especially with respect to phosphorus. CFD handled simulations of heterotrophs or PAOs alone in a reasonable computational period, typically about six hours (using a 1 GHz Intel Pentium IV Processor with 500 MB RAM). The model still converged on a solution when the equations for both organisms were included, so that they competed against each other. However, under these conditions the computations took days to complete when near steady state values were provided as the initial conditions; when initial conditions were not near steady state, a solution was not complete even after weeks of computing time.

The CFD model predicted that either heterotrophs or PAOs alone could grow in the simulated reactor. These results were consistent with those predicted by ASM2, and, for heterotrophs, an analytical solution.

However, unlike the ASM2 predictions, the CFD model indicated that heterotrophs and PAOs could coexist in a single tank over a range of oxygen inputs. This was possible because there was a second "spiral" current along with the main recirculating flow, creating a varied DO environment. The CFD results also demonstrated that oxygen input to a reactor is an important control factor even when DO concentrations are below normal detection limits.

This work is a step towards the development of a mathematical model for realistic closed loop bioreactors. Conceptually, CFD appears capable of predicting the outcomes of the process tank fluid flow pattern and biochemical reactions within realistic systems, but there is still a practical limitation regarding model complexity and computing resources.

ES.6 Recommendations

CFD has the potential to provide quantitative information about full-scale biological wastewater treatment systems. However, its usefulness is limited at present by the model complexity, cost, and computational requirements. For now these problems may limit CFD to research applications. As the software and computing systems improve, however, more routine use of CFD will become feasible. This may include opportunities to use CFD for troubleshooting at existing facilities as for designing bioreactors.

Incremental steps that can be taken to enhance the CFD model include incorporating return sludge, nitrification and denitrification reactions, the inner channels, and perhaps the effects of diffusion limitations within the biological floc. It should also be possible to refine the method of incorporating the momentum imparted by the aerators.

One exciting prospect that should be explored in future work is the possibility of designing or operating bioreactors to intentionally control DO gradients within a tank to optimize treatment. CFD can potentially play a major role in this effort. Variables may include tank shape and dimensions, and aerator rpm and submergence. Another area to explore would be the incorporation of simple baffles to promote spiral flow or otherwise increase the DO gradients created. This has the potential to enhance nutrient removal very economically in existing facilities.

CHAPTER 1.0

INTRODUCTION

1.1 Simultaneous Biological Nutrient Removal

Activated sludge is a suspended growth aerobic biological wastewater treatment process for biodegrading organic matter and converting organic nitrogen to ammonium. It also can be used to transform ammonium-nitrogen to nitrate-nitrogen via the biological process of nitrification. Additionally it can be configured to remove the nutrients nitrogen and phosphorus by providing mixed, nonaerated zones and internal process recycle streams to create the anoxic and anaerobic environments needed for biological nitrogen and phosphorus removal, respectively. Grady et al. (1999) define the functions occurring in each zone, and which ones are required to achieve different degrees of nutrient removal. Biological nutrient removal (BNR) activated sludge systems using these well-defined anoxic and anaerobic zones have evolved since the 1970s and is now a widely applied wastewater treatment technology.

However, nutrient removal beyond that needed for biomass synthesis has been observed numerous times in activated sludge facilities that do not possess explicitly defined anoxic and anaerobic zones. Nitrogen losses from aerated facilities have been observed frequently (van Munch et al., 1996; Albertson and Coughenour, 1995; van Huyssteen et al., 1990; Rittman and Langeland, 1985; Applegate et al., 1980; Drews and Greef, 1973). This phenomenon has been referred to as simultaneous nitrification and denitrification, since it was surmised that these two biological processes were occurring together in the same aerated bioreactor. Biological phosphorus removal has also been observed in aerated bioreactors with no formal anaerobic zone (Cinar et al., 1998; Brewer et al., 1995). Effluent concentrations of total nitrogen below 5 mg/L and total phosphorus below 1.0 mg/L seem to be achieved regularly at some plants, and even lower values have been reported. Simultaneous nitrogen and/or phosphorus removal (referred to hereafter as simultaneous biological nutrient removal, or SBNR) offers the potential to reduce the cost and simplify the implementation of BNR, as illustrated in Table 1-1. However, because the mechanisms responsible for SBNR are not well understood, it is not a simple matter to design and operate such systems reliably. If the mechanisms can be better characterized, SBNR could be used in a wider range of applications, thereby reducing the cost of implementing BNR and making it possible to implement it more easily at existing facilities.

Table 1-1. Advantages and Disadvantages of Simultaneous Biological Nutrient Removal (SBNR).

Advantages
• Does not require baffles to create anoxic and/or anaerobic zones. • Does not require mixing of separate anoxic and/or anaerobic zones. • Does not require mixed liquor recycle not required. • Can be implemented at some existing facilities without construction. • Reduces aeration requirements for nitrification. • Reduces alkalinity requirements for nitrification.
Disadvantages
• Operating mechanisms not well understood, so, not clear how to implement. • Process control may be more complex.

1.2 Hypothesis and Objectives

This work was conducted in the context of an overall research project directed at better understanding and controlling SBNR in full-scale wastewater treatment plants. The basic hypothesis of the overall research is that three general mechanisms may be responsible for SBNR, and that all three can operate, to different degrees, within a BNR system. The three proposed mechanisms are: 1) a bioreactor macroenvironment characterized by mixing patterns that create the anoxic and/or anaerobic zones necessary for BNR to develop informally; 2) the development of anoxic and/or anaerobic zones within the floc microenvironment; and 3) the presence of novel microorganisms with non-traditional biochemical activities. System design and operating parameters will determine the relative importance of each mechanism. The objective of the overall research project has been to identify the factors affecting the relative contributions of the three mechanisms to SBNR, thereby allowing SBNR to be implemented and controlled more effectively in full-scale plants.

As background, this report provides the results of preliminary field studies at full-scale Orbal™ plants (Section 2.2), and describes the development and application of batch assays (Section 2.3) that found the activity of novel microorganisms to be, at most, of minor importance in SBNR. It then focuses on the development and application of a computational fluid dynamics (CFD) model for closed-loop reactors to elucidate the role of the bioreactor macroenvironment in SBNR, particularly for phosphorus removal.

CHAPTER 2.0

BACKGROUND

2.1 Introduction

It is well known that many full-scale activated sludge treatment plants do not provide a uniform bioreactor macroenvironment throughout. Examples of such systems include oxidation ditches and other plants with oxygen transfer devices (such as most mechanical surface aerators) that cause large-scale recirculation of the mixed liquor (Grady et al., 1999; Cinar et al., 1998; Randall et al., 1992; van Huyssteen et al., 1990; Rittman and Langeland, 1985; Applegate et al., 1980; Drews and Greef, 1973). In such facilities, intense oxygen transfer occurs in one portion of the bioreactor, limited oxygen transfer occurs throughout the rest of the bioreactor, and mixed liquor is recycled between the aerated and nonaerated zones. This bioreactor flow pattern is also characteristic of BNR systems with defined anoxic, anaerobic, and aerobic zones. Thus, it is not surprising that SBNR would be observed in such systems.

It also is recognized that dissolved oxygen (DO) concentration gradients occur within the (AB asks, is "the" correct?)biological floc microenvironment (van Niekirk et al., 1988; Rittman and Langeland, 1985; Lau et al., 1984). Using microelectrodes to directly measure DO and oxidation-reduction potential (ORP) in activated sludge floc, Li and Bishop (2001; 2003) demonstrated that substrate and DO concentrations can contribute to stratification of microbial activities inside floc. It is possible that the anoxic and/or anaerobic zones necessary to achieve BNR form inside the biological flocs. Thus, SBNR could be observed in bioreactors where formal, identifiable anoxic and/or anaerobic zones do not form (Smith, 1996; Albertson and Coughenour, 1995; and Albertson and Stensel, 1994). In fact, the performance of an SBNR process accomplishing nitrogen removal has been characterized mathematically based upon such a mechanism (van Munch et al., 1996).

Over the past several years, a number of "new" biochemical pathways have been identified in novel microorganisms (bacteria employing these previously unrecognized pathways) that may, at least partially, account for SBNR. For nitrogen removal, these pathways include denitrification by autotrophic nitrifiers (e.g., the Anammox process) and heterotrophic nitrification – aerobic denitrification. Recent reviews have been provided by Jetten et al. (2002) and van Loosdrecht and Jetten (1998), and more detail will be included in Section 2.3. Understanding of the biochemistry and microbiology of the biological phosphorus removal process is also incomplete and evolving (Mino et al., 1998). Thus, future discoveries may reveal currently unknown organisms that can achieve enhanced biological phosphorus removal in aerated bioreactors.

It is the basic hypothesis of this SBNR research program that all three of these mechanisms (bioreactor macroenvironment, floc microenvironment, and novel microorganisms) might operate in systems accomplishing SBNR. However, it is logical that the relative contributions of each mechanism may vary, depending, for example, upon system design and operating parameters. Thus, a key to understanding and controlling SBNR is to understand how process design and operating parameters affect each of these three mechanisms. This is the basic objective of the overall SBNR research program.

2.2 Preliminary Field Study

Before starting the WERF project, a preliminary field study was conducted (Daigger and Littleton, 1999, 2000) to characterize SBNR in some representative full-scale activated sludge plants that are either known or thought to be capable of SBNR. Seven plants were evaluated to obtain an overall assessment of performance capabilities (Table 2-1). Selected plants were then evaluated in more detail.

Table 2-1. Summary of Wastewater Treatment Plants Studied.

Plant, Location	Design capacity	Average flow	Organic load	Trains	Design HRT	SRT	MLSS	Operating Period
	(m^3/day)	(% design)	(% design)	(no.)	(hr)	(days)	(mg/L)	(dates)
Elmwood, Evesham, NJ[1, 2]	11,400	63	55	2	22.6	33	3,175	1-9/98
Hartford Road, Mount Laurel, NJ	22,700	66	47	1	14.5	30	3,500	1-9/98
Hammonton, NJ[1, 2]	9,500	36	46	1	19.0	20	2,200	7/94-6/95
		39	41			26	2,400	1/98-3/99
Chalfont, New Britain, PA	15,100	75	91	1	14.0	24	4,000	1-12/94
		101	132			8	2,752	2/98-1/99
McMinnville, OR	21,210	100	61	2	12.9	11	2,146	1-12/97
Sweetwater Creek, Gwinnett County, GA[1]	45,400	89	85	4	11.0	7-10	3,411	1-12/94
Lake Geneva, MI[2]	6,600	87	115	1	15.3	22	4,000	1-12/94

[1] Plant includes effluent filtration. [2] Mixed liquor recirculation from Channel 3 to Channel 1.

2.2.1 Orbal™ Plants

The focus was on plants of the Orbal™ configuration, in which the process bioreactor generally consists of three concentric aerated closed-loop reactors in series (Figure 2-1). Influent and return sludge are added to the outer channel (Channel 1), then flow to the middle (Channel 2) and inner (Channel 3) stages before entering a secondary settling tank. Oxygen is transferred using disc aerators, and oxygen input to each stage is varied to allow the creation of different environmental conditions (Smith, 1996). The aerators also provide mixed liquor recirculation within each channel to maintain a sufficient velocity (generally $\geq$0.3 m/sec) to keep settleable solids in suspension. The result is that the mixed liquor is recirculated around the loop every few minutes, while the average hydraulic residence time in each channel will be several hours. As a result, spatially varying environments can be created through (AB asks: through or throughout?) the bioreactor, thereby allowing the first mechanism described above (macroenvironment mixing patterns) to operate. Moreover, because all three bioreactor stages are aerated and oxygen input can be independently controlled, the potential exists to develop anoxic and/or anaerobic microenvironments within the biological floc (second mechanism). This makes the Orbal^{TM} configuration attractive for the study of SBNR. Moreover, focusing on plants with one specific configuration made it easier to characterize the effects of other design and operating parameters, such as solids retention time (SRT), hydraulic residence time (HRT), and oxygen input. The specific objectives of this study were to determine 1) how much simultaneous nitrogen and/or phosphorus removal was occurring in selected full-scale Orbal^{TM} plants, and 2) how process operating parameters affect the extent of SBNR.

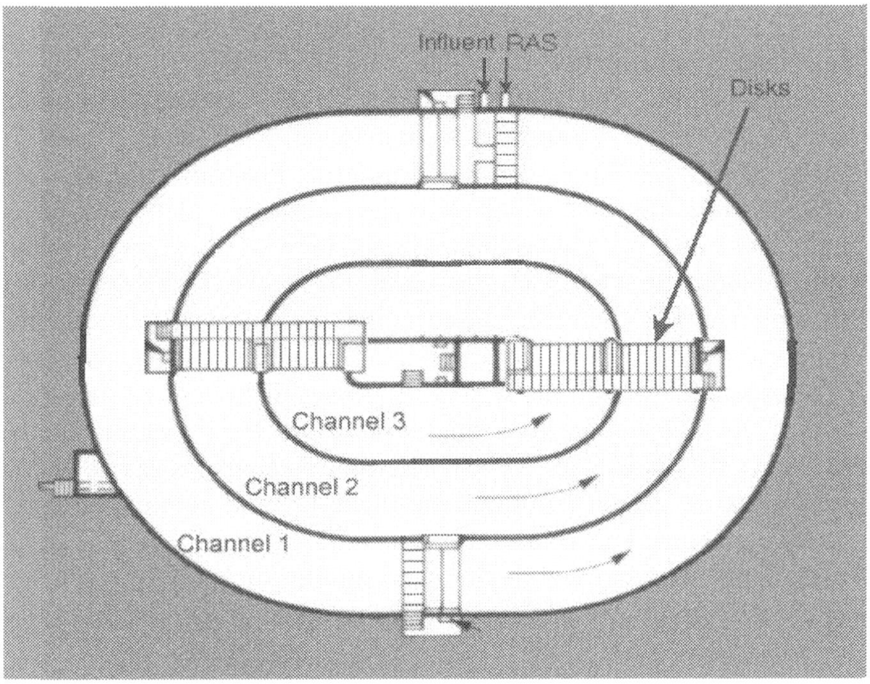

Figure 2-1. Orbal^{TM} Process Schematic.

A typical design and operating strategy for the Orbal^{TM} process is to provide less oxygen to Channel 1 than required to meet the full process oxygen demand (Smith, 1996). Typical calculations indicate that about 70% of the total process oxygen demand will be exerted in Channel 1. However, the oxygen supply provided in Channel 1 is 50-70% of the calculated demand. This is done to allow both nitrification and denitrification to occur in the outer channel. Oxygen provided to the channel allows nitrification to occur, but because oxygen input is restricted, the nitrate-nitrogen produced through nitrification is consumed. It is presumed that heterotrophic microorganisms denitrify the nitrate-nitrogen produced by autotrophic nitrifiers. This approach also seems to help control low DO bulking filaments that might compete well at slightly higher DO concentrations. Grady et al. (1999) and Randall et al. (1992) discuss manipulation of oxygen supply to control nitrification and denitrification. Mixed liquor can also be recycled from Channel 3 to Channel 1 to transport nitrate-nitrogen formed in Channels 2 and 3 back to Channel 1 where it can be denitrified. Use of these strategies can remove more than 90% of the nitrogen (Smith, 1996; Applegate et al., 1980; Drews and Greef, 1973).

2.2.2 Analytical Methods

Much of the data used were collected as part of the normal operation and monitoring of the plants. Influent and effluent data were from 24-hour flow-proportioned composite samples. The sampling frequency in most cases ranged from two to five times per week. Grab sample data were available to evaluate the variation in dissolved parameters throughout the bioreactor, including DO. Tests were generally conducted in accordance with *Standard Methods for the Examination of Water and Wastewater* (Clesceri et al., 1998) or equivalent procedures (U.S. EPA, 1983), in laboratories that were certified for the parameters reported. Typically, authorities operating the wastewater treatment plants conducted the analyses as a part of their normal process control and/or permit compliance monitoring.

2.2.3 Process Modeling Techniques

An analysis of the operating characteristics of the Orbal™ process was conducted using the International Water Association's Activated Sludge Model Number 1 (ASM1) (Henze et al., 1987). It was implemented with the Single Sludge Simulation Program (SSSP) developed by Bistrup and Grady (1988) using the wastewater characterization and plant simulation tool called PRO^{2}D (Daigger et al., 1998). ASM1 was used to determine to what extent simultaneous nitrification and denitrification could be characterized in this process using generally accepted process models. This model was selected for two principal reasons. First, it is the consensus suspended growth biological process simulation model (Grady et al., 1999). Second, it can characterize SBNR in suspended growth biological processes that occur as a result of the first two SBNR mechanisms hypothesized in this research program: 1) bioreactor macroenvironment and 2) floc microenvironment (Cinar et al., 1998). The first mechanism can be incorporated by configuring the bioreactor at the proper combination of aerated and nonaerated zones with mixed liquor recirculation between them. The second mechanism is included because of the differences in half saturation coefficients for dissolved oxygen and nitrate-nitrogen for the heterotrophic bacteria and because the half saturation coefficient for dissolved oxygen is lower for heterotrophs than for autotrophs.

2.2.4 General Process Results

Table 2-2 summarizes overall plant performance. Removal of five-day biochemical oxygen demand (BOD_5) and total suspended solids (TSS) was excellent. At some of the plants, this includes effluent filtration; however, biological process effluent concentrations were not much different from filtered effluent values in these cases because very few suspended solids were generally present in the biological process effluent to be removed by the filters. Nitrification also was essentially complete, with effluent ammonium-nitrogen (NH_4^+-N) values often <1 mg/L. Although nitrification was extensive, effluent nitrate-nitrogen (NO_3^--N) values were generally less than 5 mg/L, often much less. Yearly average effluent NO_3^--N values as low as 1-3 mg/L are recorded at some plants, which is surprisingly low. At the Elmwood and Hammonton plants, where influent total Kjeldahl nitrogen (TKN) data were available, total nitrogen removals were 85-90%. The average effluent total nitrogen concentration (TKN + nitrate-N) for the Lake Geneva plant was <4 mg/L, and it was <6 mg/L at McMinnville, OR. This indicates that substantial total nitrogen removal was occurring at each facility. The few plants with effluent total nitrogen discharge limits were all in compliance.

Table 2-2. Summary of Plant Performance Data.

Plant	BOD_5 (mg/L)		TSS (mg/L)		TPO_4^{3-}-P (mg/L)		TKN (mg/L)		NH_4^+-N (mg/L)		NO_3^--N (mg/L)
	In	Out	In	Out	In	Out	In	Out	In	Out	Out
Elmwood	221	2.3	184	1.1	5.4	0.5	32.5	2.0	25.0	1.1	1.1
Hartford Road	210	3.6	292	4.8	-	-	-	-	-	0.1	-
Hammonton - 1994/95	353	2.1	390	4.2	-	1.7	37.0	2.1	-	0.2	2.9
- 1998/99	294	1.0	302	1.5	3.0	1.2	36.6	1.5		0.3	1.4
Chalfont - 1994	160	3.2	152	4	3.2	0.9	-	-	15.8	1.0	5.5
- 1998/99	172	4.1	158	6.5	3.6	1.4			20.6	0.8	6.7
McMinnville	117	2.4	192	4.6	3.9	0.4		2.0		1.4	3.6
Sweetwater Creek	237	1.8	359	1.5	6.0	0.2	-	-	13.0	0.1	4.5
Lake Geneva	203	4.2	196	6.2	-	-	-	1.3	-	-	2.6

Effluent total phosphate-phosphorus (TP) data were available for five of the facilities and indicate removal to effluent concentrations of 1.7 mg/L or lower. Phosphorus removal was 0.022 mg P/mg BOD_5 at the Elmwood plant, while it was 0.024 mg P/mg BOD_5 at Sweetwater Creek. This exceeds the phosphorus removal expected for biomass synthesis alone for a long SRT process and suggests that some other mechanism is contributing to phosphorus removal. Since metal salts are not added at these two plants, enhanced biological phosphorus removal may be occurring. At McMinnville, metal salts are added during the summer, but effluent TP concentrations are below 1 mg/L year round. Phosphorus removal by the biological process averaged about 0.03 mg P/mg BOD_5. Influent total phosphate concentrations are low at some of the facilities, however, so phosphorus removal in those cases may simply be for biomass synthesis. Small amounts of ferric chloride are added to the Chalfont facility to assist with sludge settleability. Still, taken together the data suggest that enhanced biological phosphorus removal may be occurring at some of the plants. Some of these plants have effluent TP discharge limits, and all of these systems were in compliance.

2.2.5 Data Suggesting Simultaneous Removal of Nitrogen

DO concentrations in the bioreactor at the Elmwood plant were measured during the more highly loaded period of the day with a portable DO meter (YSI Incorporated, Yellow Springs, OH). Little difference was found between measurements made 1.5 m upstream or downstream of the aerators. DO concentrations were low, typically ≤0.2 mg/L in Channel 1, 0.2-0.4 mg/L in Channel 2, and 0.2-2 mg/L in Channel 3. These results are also typical for several of the other facilities evaluated and comparable to those obtained by Applegate et al. (1980) in their study of the Orbal™ system in Huntsville, TX. These data indicate that distinct aerobic and anoxic zones, as indicated by high and low DO concentrations, do not necessarily develop within the Orbal™ bioreactor. In contrast, the Chalfont plant, which has less stringent effluent total nitrogen limits, operates at higher average DO concentrations. The DO profiles for this plant indicated a vertical DO profile (decreasing with depth), especially in Channel 1. This suggests some vertical stratification within the channel of soluble constituents (but not of the mixed liquor suspended solids, MLSS, concentration).

Grab samples of the influent wastewater, return activated sludge (RAS), and the mixed liquor were collected periodically during the daytime at the Elmwood plant and filtered immediately, and the nutrient concentrations were then measured to characterize their concentrations through the bioreactor. Some typical data are presented in Table 2-3, along with some additional DO profile data. The calculated influent is the flow weighted average of the biological process influent and the RAS, which is the concentration expected in Channel 1 if no reactions occurred there. The results indicate that, even when substantial nitrification occurs in Channel 1, as indicated by low soluble TKN and ammonium-nitrogen concentrations, the nitrite- and nitrate-nitrogen concentrations do not increase appreciably. This suggests that nitrification and denitrification are both occurring in Channel 1.

Table 2-3. Example Elmwood Plant Nutrient Profile Data (mg/L).

Location	TKN[1]	TPO_4^{3-}-P[1]	NH_4^+-N	NO_3^--N	NO_2^--N	DO
4/17/96						
Calculated Influent	12.0	0.91	8.8	-	-	-
Orbal ™ 1						
Channel 1	3.1	0.55	2.2	<0.4	0.02	-
Channel 2	1.1	0.34	1.3	1.3	0.10	-
Orbal ™ 2						
Channel 1	3.6	1.03	2.7	0.8	<0.02	-
Channel 2	<1.0	0.53	1.6	0.9	0.02	-
4/15/99						
Calculated Influent	6.5	0.6	4.4	3.0	-	-
Orbal ™ 1						
Channel 1	5.85	0.48	4.8	<0.4	-	0.08
Channel 2	2.61	0.05	1.7	1.42	-	0.18
Channel 3	1.28	0.06	<0.1	3.09	-	0.81
Orbal ™ 2						
Channel 1	5.62	1.06	5.0	<0.4	-	0.07
Channel 2	2.60	0.54	2.2	<0.4	-	0.28
Channel 3	1.05	0.13	<0.1	1.09	-	1.70

[1] Measured after filtration through glass fiber filter.

ASM1 was used to characterize nitrogen removal in the Elmwood plant (Table 2-4). In the simplified model, Channel 1 was simulated as six equal-sized, completely mixed cells in series and Channels 2 and 3 were simulated as individual completely mixed cells. In Channel 1, aeration zones in two cells directly opposite each other represented the locations of the disc aerators, while the other four cells were non-aerated. Channels 2 and 3 were aerated. A range of oxygen mass transfer coefficients (K_La values) was input to the simulation, and the resulting DO concentrations calculated. This approach was selected because it did not require DO concentrations to be specified in each zone, and because it allowed for non-zero DO concentrations even in the non-aerated zones (from the inflow of oxygenated mixed liquor from an upstream aerated zone). To simulate the mixed liquor recirculation created by the action of the disc aerators, the Channel 1internal recirculation flow rate was 580,000 m^3/day (sufficient to provide a velocity of 0.3 m/sec). In some scenarios, a mixed liquor recirculation pumping rate of 22,800 m^3/day from Channel 3 to Channel 1 was also provided. The simulations also used average process and wastewater characteristics (Tables 2-1 and 2-2) and a temperature of 20ºC. The calculated MLSS concentration was 3,117 mg/L, which corresponds well with the reported value of 3,175 mg/L (Table 2-1).

Table 2-4. Summary of Process Analysis of Elmwood Plant Using ASM1.

Location	Volume (m^3)	Concentration (mg/L)								
		Excess Oxygen			Oxygen Limited			Oxygen Limited[1]		
		DO	NH_4^+-N	NO_3^--N	DO	NH_4^+-N	NO_3^--N	DO	NH_4^+-N	NO_3^--N
Channel 1[2]	5,375									
Cell 1	895	2.9	0.5	19.7	0.6	1.4	2.1	0.6	1.0	4.0
Cell 2	896	2.3	0.4	19.7	0.3	1.3	2.0	0.4	1.0	3.9
Cell 3	896	1.9	0.4	19.7	0.1	1.3	2.0	0.2	0.9	3.8
Cell 4	896	2.9	0.3	19.7	0.6	1.2	2.0	0.7	0.9	3.8
Cell 5	896	2.5	0.3	19.8	0.3	1.2	2.0	0.4	0.8	3.8
Cell 6	896	2.1	0.2	19.8	0.2	1.2	1.9	0.2	0.8	3.8
Channel 2	3,560	2.4	0.1	19.5	2.2	0.1	1.8	1.6	0.1	3.6
Channel 3	1,815	2.7	0.1	19.6	2.8	0.1	1.8	2.3	0.1	3.4

[1] Mixed liquor recirculation from Channel 3 to Channel 1 of 22,500 m^3/day.

[2] Internal recirculation (Cell 6 to Cell 1) rate to achieve 0.3 m/sec channel velocity = 580,000 m^3/day; Cells 1 and 4 have aerators.

Three operating scenarios were evaluated (Table 2-4). The first provided sufficient oxygen input into Channels 1, 2, and 3 to maintain DO concentrations ≥2 mg/L. Nitrification was predicted to be essentially complete in Channel 1, with little nitrification and nitrate production in Channels 2 and 3. This is a result of the relatively long operating SRT (33 days). For this simulation, nearly 75% of the total process oxygen demand (2,250 kg/day) was exerted in Channel 1, 20% in Channel 2, and the remainder in Channel 3.

The second simulation restricted the oxygen input to Channel 1 to allow nitrification and denitrification to occur there. The lowest effluent total nitrogen concentration was observed when about 50% of the oxygen was added to Channel 1, 35% to Channel 2, and 15% to Channel 3. The total process oxygen requirement was reduced 24%, to 1,710 kg/day. The predicted DO and nitrogen concentrations (Table 2-4) were similar to those observed in the full-scale plant (see Tables 2-2 and 2-3). Sufficient oxygen was supplied in Channel 1 so that most of the ammonium was nitrified there, but restricting the oxygen input also allowed substantial denitrification to occur.

The third simulation recycled mixed liquor from Channel 3 to Channel 1 . Surprisingly, effluent nitrate-nitrogen concentrations increased slightly. This occurred because recirculating mixed liquor from Channel 3, which contained low organic matter, reduced the concentration of the organic matter available for denitrification in Channel 1. This is not a generalizable result, but it does illustrate the interactive nature of these systems.

It is emphasized that all three simulations used default kinetic and stoichiometric parameters and procedures for translating conventional wastewater characteristic data into the specific constituents required for ASM1 (Henze et al., 1987; Bistrup and Grady, 1988; Daigger et al., 1998). The fit of the model could be improved by adjusting these parameters. The fact that a good fit was obtained with default values and procedures suggests, however, that the model may provide a reasonable representation of the system.

In a separate set of simulations (results not shown), the process was characterized as only three completely mixed cells in series. Thus, Channel 1 was assumed to function as a single, completely mixed cell. Again, the researchers predicted complete nitrification, and limited or no denitrification when DO concentrations greater than 2 mg/L were maintained in all three cells. They predicted substantial denitrification when the DO concentrations in Channels 1 and 2 were reduced, while the process effluent was still fully nitrified. As might be expected, these results did not match actual values as well as did the more complex model presented in Table 2-4. However, they further demonstrate that creating formal anoxic zones is not necessary to allow ASM1 to predict significant nitrogen removal in staged systems such as Orbal™.

These results suggest that nitrogen removal can be characterized using existing process models with default kinetic and stoichiometric parameters and default analysis procedures. The simulation results presented in Table 2-4 do not predict distinct aerobic and anoxic zones, due to the high mixed liquor recirculation rate in Channel 1 created by the pumping action of the aerator necessary to maintain settleable solids in suspension. In fact, denitrification was still predicted even when each channel is assumed to be a completely mixed cell. Thus, the second SBNR mechanism discussed above (the floc microenvironment) appears to be sufficient to characterize (AB asks: characterize or permit?) nitrogen removal in these processes. It is also possible that the first mechanism (the bioreactor macroenvironment) contributes to nitrogen removal in these systems. Note, however, that it was not necessary to incorporate such refinements into the model for nitrogen removal developed for the Elmwood plant. ASM1 is based on: 1) independent functioning of autotrophic nitrifiers and heterotrophic denitrifiers and 2) no denitrification by autotrophs. Thus, it appears that novel microorganisms are not necessary for simultaneous nitrogen removal, although they may still be present. These results provide a conceptual basis to begin analyzing the nitrogen removal capability of these systems.

2.2.6 Data Suggesting Simultaneous Removal of Phosphorus

As discussed above, effluent total phosphate data from at least three of the plants studied suggested that enhanced biological phosphorus removal might be occurring. Two plants (Elmwood and Sweetwater Creek) had influent wastewater BOD_5/TP ratios of about 40 mg BOD_5/mg TP, while the third (McMinnville) had a ratio of 30 mg BOD_5/mg TP. These values are favorable to achieve excellent biological phosphorus removal (Grady et al., 1999). Further support for the presence of biological phosphorus removal was provided by microscopic examination of the mixed liquor from the Elmwood and Hammonton plants, which indicated the presence of bacteria containing polyphosphate granules (Jenkins, 1999). The phosphorus content of the MLSS at the Elmwood plant was measured to be about 2.5% (P/VSS), indicating an accumulation of phosphorus in excess of that associated with biomass synthesis and consistent with the occurrence of enhanced biological phosphorus removal.

As indicated by the data for the Elmwood plant presented in Table 2-3, dissolved total phosphate concentrations in Channels 1 and 2 were generally low. One might expect to see elevated phosphate concentrations in one of the channels, suggesting phosphate release associated with enhanced biological phosphorus removal (Grady et al., 1999). However, this was not the case.

Cinar et al. (1998) studied biological phosphorus removal in a closed-loop bioreactor using disk aerators of the same type used in the Orbal™ process, but they were unable to characterize it mathematically using ASM2, an extension of ASM1 that incorporates biological phosphorus removal. One problem was the high nitrate loading on the various segments of

Channel 1 due to the high mixed liquor recirculation rate. While it is possible that anaerobic zones develop within biological flocs, this is not sufficient (unlike the case with nitrogen) to explain the occurrence of biological phosphorus removal. For one thing, it is likely that the anaerobic zone would form in the center of the floc where readily biodegradable organic matter may not penetrate. Moreover, the responsible organisms would not cycle between aerobic and anaerobic conditions, as is required. Thus the floc microenvironment mechanism alone cannot explain biological phosphorus removal in these systems.

It is possible, though, that one or more anaerobic zones develop in Channel 1, perhaps in the deeper layers. The DO profile data from Chalfont suggests this. Moreover, discussions with the manufacturer of the disc aerators confirm that flow patterns in these systems consistent with this hypothesis have been observed in mixing tests (Dobbeck, et al., 1999). If the mixing pattern created by disc aerators is verified to form an aerobic zone naturally, this type of system gains a potential advantage in the ability to remove nutrients. More detailed analysis of the flow patterns in Channel 1 is necessary to evaluate this possibility. However, this provides a reasonable hypothesis concerning the occurrence of biological phosphorus removal in these systems. In fact, process simulations conducted using ASM2 have indicated that such flow patterns would result in the occurrence of biological phosphorus removal (Filipe, 1999). This result is similar to that of Cinar et al. (1998) in another plant. They found that vertical (rather than horizontal) mixing patterns could explain the removal of biological nitrogen and phosphorus in a full-scale bioreactor. This further points to the need to examine the first proposed mechanism, the reactor macroenvironment created by the aeration device.

2.2.7 Summary and Conclusions

Substantial nutrient removal was observed routinely in staged, closed-loop bioreactors. Essentially complete nitrification occurred, as well as extensive total nitrogen removal. Biological phosphorus removal was observed in some of the facilities, as evidenced by a phosphorus-to-BOD_5 removal ratio exceeding that associated simply with biomass synthesis, and by the observation of microorganisms with polyphosphate granules in the mixed liquor. Thus, while the full nutrient removal capacity of these facilities has not been quantified, the plants demonstrate that good nutrient removal can be achieved with facilities not intentionally designed to incorporate the formal anaerobic and anoxic zones required in conventional BNR plants.

Nutrients are removed in these systems by restricting the oxygen input to Channels 1 and 2, which results in low DO concentrations there. Both nitrification and denitrification occur in these channels, resulting in substantial nitrogen removal. This operating mode also selects for the organisms responsible for biological phosphorus removal. The oxygen input into these channels controls the extent of nutrient—the greater the degree of oxygen deficiency (within limits), the greater the nutrient removal.

Analysis of the operating characteristics of these facilities suggests that SBNR can be characterized using the first two proposed mechanisms, the bioreactor macroenvironment and the floc microenvironment. Each channel is relatively well mixed in a horizontal direction as a result of the high mixed liquor recirculation flow rate required to maintain settleable solids in suspension. However, there may be vertical stratification of dissolved oxygen or other soluble constituents that potentially change rapidly in concentration due to uptake or release (not mixed liquor suspended solids), so anaerobic zones could develop in portions of Channel 1 (and possibly Channel 2). Anaerobic zones may also develop within the biological flocs. Conceptually at least, formation of anoxic conditions within the flocs can explain nitrogen removal, but it cannot explain phosphorus removal. While novel microorganisms may be present in the mixed liquor, their presence is not necessary to explain the nutrient removal occurring.

2.3 Evaluation of the Role of Novel Microorganisms in the SBNR of Full-Scale Systems

Recent advances in microbiology have shown that there are previously unknown microorganisms and biochemical pathways (referred to here as novel microorganisms) that might contribute to SBNR. Some examples are shown in Table 2-5. The reviews by Jetten et al. (2002) and van Loosdrecht and Jetten (1998) provide a detailed overview of nitrogen conversions described in the literature. Littleton et al. (2003) have summarized much of this literature in relation to SBNR.

Table 2-5. Examples of Potential Novel Biochemical Conversions in SBNR Systems.

Biochemical Conversion	Biochemical Characteristics	Impacts on Process
Denitrification by Autotrophic Ammonium Oxidizers: Some typical autotrophic ammonium oxidizers (*e.g.*, *Nitrosomonas*) and planctomycetes (*Brocadia*)) can use ammonium as an electron donor and nitrite as an electron acceptor when DO is absent. The product is nitrogen gas (Jetten et al., 1999; Strous et al., 1999 a & b; Kuai & Verstraete, 1998; Bock et al., 1995).	Occurs when ammonium and nitrite are present and DO is absent.	- Carbon requirements for N removal reduced since biodegradable organic carbon not needed. - Process oxygen requirements reduced if supplemental carbon addition is reduced.
Heterotrophic Nitrification – Aerobic Denitrification: Some heterotrophs (*e.g.*, *Paracoccus pantotropha*) are able to oxidize ammonium to nitrite using pathway similar to *Nitrosomonas* (Robertson and Kuenen, 1990; Geraats et al., 1990).	Nitrification consumes energy (rather than providing it) for these organisms and requires biodegradable organic material.	- May expand range of environmental conditions (SRT, DO, pH) over which nitrification occurs. - May result in reduced process yield (sludge production).
Aerobic Denitrification: Some heterotrophs are able to denitrify even when DO is present (Robertson and Kuenen, 1984).	Simultaneous use of oxygen and nitrate as terminal electron acceptors allows higher specific growth rate.	- Expanded range of environmental conditions (DO) over which denitrification occurs.
Denitrification by PAOs: Some PAOs are able to use either DO or nitrate as a terminal electron acceptor (Filipe and Daigger, 1997 and 1999; Kuba et al., 1997).	Conventional enhanced biological phosphorus removal metabolism, but using nitrate as terminal electron acceptor.	- Carbon requirements for nutrient removal reduced as carbon used by denitrifying PAOs gives both P removal and denitrification

A study (Littleton et al., 2000, 2002a, 2003a) was begun prior to this WERF project to examine the importance of such microorganisms in full-scale plants. The researchers further investigated three of the previously characterized (Section 2.2) Orbal™ plants using a screening tool developed to determine the relative significance of the activities of microorganisms capable of autotrophic denitrification and heterotrophic nitrification/aerobic denitrification in biological nutrient removal systems. In one of the plants, they also tested phosphorus metabolism (anaerobic release, aerobic uptake).

2.3.1 Novel Biochemical Conversions of Interest

2.3.1.1 Denitrification by Autotrophic Ammonium Oxidizers

It has been observed that, when ammonium and nitrite are present and DO is absent, some bacteria can use ammonium as the electron donor and nitrite as the electron acceptor. This type of autotrophic denitrification results (Equation 2.3.1.1-1) in conversion of both nitrogen species to nitrogen gas (Jetten et al., 1999; Strous et al., 1999a; van Loosdrecht and Jetten, 1998; Kuai and Verstraete, 1998; Bock et al., 1995). (Note that there are also autotrophic organisms that oxidize reduced sulfur or other inorganic substances that are capable of denitrification, but the term autotrophic denitrification will be used here for nitrifying denitrifiers only.)

$$NH_4^+ + NO_2^- \rightarrow N_2 + 2H_2O \qquad (2.3.1.1\text{-}1)$$

This reaction has been employed commercially in the Anammox process (Jetten et al., 2002; Strous et al., 1997; Bock et al., 1995). The diversity of microorganisms catalyzing this transformation has not been fully explored but includes some species of *Nitrosomonas* as well as a newly identified member of the planctomycetes, *Brocadia* (Jetten et al., 1999; Strous et al., 1999b; Madigan et al., 2003). Anammox organisms grow extremely slowly (Jetten et al., 1999; Strous et al., 1999a and b; Kuai & Verstraete, 1998; Van Loosdrecht & Jetten, 1998), and this reaction has only been observed in the absence of available organic substances. However, the SRTs used in BNR systems and the low DO environment that may occur there might allow this reaction to take place.

2.3.1.2 Heterotrophic Nitrification/Aerobic Denitrification

Some heterotrophs, such as *Paracoccus (*previously *Thiosphaera) pantotropha*, can aerobically convert ammonium to nitrite using biochemical pathways similar to those used by the autotrophic nitrifiers. However, these reactions require rather than provide energy (Robertson et al., 1988). These organisms are also capable of both aerobic and anaerobic denitrification. Heterotrophic nitrification and aerobic denitrification by *P. pantotropha* during growth on organic substrates such as acetate may be explained (Robertson et al., 1988, Geraats et al., 1990; Robertson and Kuenen, 1990) by a bottleneck in electron transfer to O_2 (Figure 2-2).Under either aerobic or anoxic conditions, when electron transfer to oxygen is rate limiting, the excess electrons flow to nitrate or nitrite (denitrification) if it is available.. If nitrate and nitrite are both absent, but oxygen is present, energy is used to oxidize ammonium (heterotrophic nitrification), making denitrification possible. This reaction occurs when both ammonium and readily biodegradable organic matter (typically acetate) are present under aerobic conditions. Since heterotrophic nitrification requires energy, there is reduced yield in pure culture. The specific nitrifying activity of heterotrophs is lower than that of autotrophs. However, in competitive systems where these conditions may exist, such as BNR treatment plants, there may still be a competitive advantage for such organisms compared to other heterotrophs.

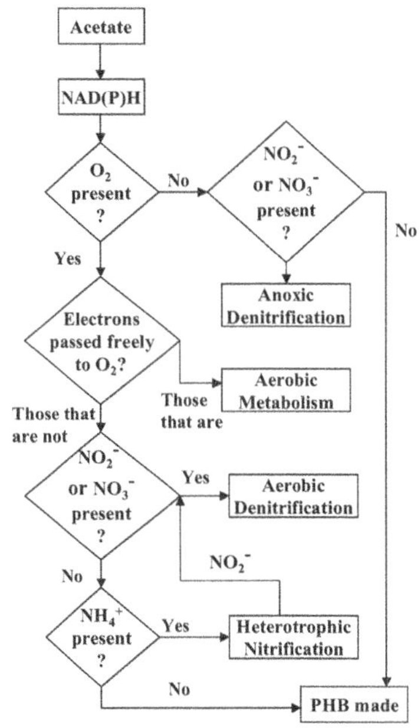

Figure 2-2. *Paraccocus pantropha* Metabolism under Varying Environmental Conditions. (Based on Robertson and Kuenen, 1990; reprinted with permission from Littleton et al., 2003a.)

2.3.2 Objectives

The primary objective of this phase of the overall research project was to use simple batch assays to determine the principal biochemical activities responsible for BNR, and especially SBNR. Using the known biochemistry of these organisms, the researchers developed procedures to distinguish the relative contributions to observed SBNR activity of the novel biochemistry of autotrophic denitrification versus traditional heterotrophic denitrification, and of heterotrophic nitrification versus traditional autotrophic nitrification. The procedures were then used as a screening tool to determine whether the activity of such organisms great enough to represent a substantial SBNR mechanism that must be considered in the analysis of full-scale systems.

The test procedure focused on determining the activity of the novel microorganisms, not simply their presence or absence, by assessing how quickly these organisms use the various nitrogen species when appropriate conditions are provided. The researchers assessed the magnitude of the contribution of these organisms relative to the conventional biochemical processes of autotrophic nitrification and heterotrophic denitrification. A second objective was to determine whether the activity of PAOs conformed to existing biochemical models. Thus, the specific objectives were to:

1) develop a simple and effective assay method to determine whether the activities of autotrophic denitrifiers and heterotrophic nitrifiers are significant in full-scale systems;
2) apply this method in full-scale plants to determine whether the activity of autotrophic denitrifiers and heterotrophic nitrifiers contributes substantially to SBNR; and

3) determine whether the activity of PAOs was consistent with current knowledge (i.e., anaerobic phosphorus release, aerobic phosphorus uptake).

2.3.3 Materials and Methods

2.3.3.1 Nitrogen Removal Batch Assays

Mixed liquor from a wastewater treatment plant achieving SBNR was removed and subjected to a series of batch experiments. The researchers used a factorial experimental design was used whereby the conditions were varied either to preclude or to emphasize the activity of the novel or traditional microorganisms in the existing mixed liquor, based on their known biochemistry. The maximum potential activity of each of these organisms was then determined by comparing the relative activities measured in the groups of reactors with appropriate controls. The researchers assayed autotrophic denitrification activity using anoxic batch reactors, and assayed heterotrophic nitrification activity using aerobic batch reactors.

As described above, autotrophic denitrification requires the presence of both ammonium and nitrite in the absence of dissolved oxygen. This activity was assayed under anoxic conditions using a 2 by 2 factorial experimental design (Table 2-6) with the two variables being the presence or absence of ammonium and nitrite. It was necessary to add nitrate to all four reactors to minimizenitrite consumption by heterotrophic denitrifying bacteria, which will undergo endogenous respiration under these conditions. Reactor 4, with both ammonium and nitrite, allows autotrophic denitrification to proceed at its maximum potential rate. Reactor 3, with ammonium but without nitrite, provides a control for Reactor 4 for ammonium production, for example by the decay of heterotrophs present. Likewise, Reactor 2, with nitrite but without ammonium, provides a control for Reactor 4 regarding nitrite reduction by heterotrophs through endogenous respiration. Reactor 1 serves as an unfed anoxic control.

Table 2-6. Batch Assay Experimental Design.

Test for	Autotrophic Denitrification		Heterotrophic Nitrification	
Assay	Anoxic		Aerobic	
Ammonium Added	Nitrite Added		Acetate Added	
	No	Yes	No	Yes
No	Reactor 1	Reactor 2	Reactor 5	Reactor 6
Yes	Reactor 3	Reactor 4	Reactor 7	Reactor 8

Heterotrophic nitrification requires the presence of ammonium and readily biodegradable organic matter (such as acetate) under aerobic conditions. Again, a 2 by 2 factorial **(AB: hyphens?)** experimental design was used, but under aerobic conditions (Table 2-6). Reactor 8, with both ammonium and acetic acid added, allows heterotrophic nitrification to proceed at the maximum rate. Reactor 7, with ammonium but no acetic acid, provides a control for Reactor 8 for nitrification by autotrophic nitrifiers. Reactor 6, with acetic acid present but no ammonium, provides a control relative to the consumption of ammonium for heterotrophic growth. Reactor 5 is the unfed aerobic control.

To evaluate several microbial communities, field experiments were conducted using mixed liquor from the Elmwood, Hartford Road, and Hammonton plants described in Section 2.2

(Tables 2-1 and 2-2). All are Orbal™ oxidation ditches that accomplish significant SBNR (Daigger and Littleton, 2000). The assays were conducted using fresh mixed liquor straight from the third channel of the full-scale plants, with modest amounts (~20 mg/L ammonium-, nitrite-, and nitrate-N; ~80 mg/L as COD of potassium acetate) of appropriate added substrate, and a relatively short reaction time. compared to the doubling times of the relevant microorganisms. This time scale limited growth so that the assays determined the activity of the organisms present in the existing mixed liquor.

The test bioreactors were sealed with a gasketed stainless steel lid and operated using an initial working liquid volume of 1 L and a headspace volume of 0.5 L. All experiments were conducted at room temperature (21-22°C). Dissolved oxygen (DO) and pH were monitored, and 4.4 mL of stock KH_2PO_4 buffer solution (pH 7.2) was added to all reactors (final concentration 150 mg/L) to minimize pH changes. Magnetic stirrers were used for mixing. For the anoxic tests, samples (50 to 75 mL) were collected twice in the first hour and then at one-hour intervals for six to eight hours.For the aerobic tests, samples were collected at 15 and 30 minutes and then every half hour for 3.5 hours.. They were immediately centrifuged to remove biological solids, then analyzed for inorganic nitrogen species (ammonium, nitrite, and nitrate) in accordance with *Standard Methods* (Clesceri et al., 1998) in the laboratory at the Elmwood plant.

Anoxic reactors were purged with N_2 gas via spargers before the test solution was added and during each sampling event. A three-way needle valve was used to release purged nitrogen off-gas, introduce test solution, and sample. The valve remained in the closed position during reaction. A positive gas pressure was maintained with nitrogen gas in the reactors in order to prevent leakage of oxygen into the reactors, especially during the intensive sampling periods. Samples were drawn carefully to maintain this positive pressure. After sampling, nitrogen gas was continuously purged for several minutes, then the three-way needle valve was switched back to the closed position to prevent oxygen from being transferred into the anoxic reactors. When reactor headspace volume had increased to 800 mL due to repeated sampling, nitrogen gas was purged continuously to the end of the test period.

The length of the anoxic test period was based approximately on the volumetric nitrogen reduction rate observed in the outer channel of the Orbal™ plant. For instance, if the plant total amount of nitrogen removal in the outer channel was 1.4 mg/L-hr and the ammonium-nitrogen added in the anoxic batch reactor was 20 mg/L, then a seven-hour test period was used to allow the researchers to observe a comparable (50%) nitrogen reduction. In other words, if substantial autotrophic denitrification was not observed within this period, then this activity was not contributing substantially to the nitrogen removal occurring in the full-scale plant.

No nitrate was added to the aerobic batch reactors, and aeration with compressed air was initiated via the sparger to maintain DO concentrations greater than 4 mg/L. These tests were conducted for 3.5 hours. No special precautions were necessary concerning sample withdrawal since the reactors were continuously sparged with air. (Additional experimental details are available in Littleton et al., 2003a.)

With the anoxic assays, the researchers determined evidence of autotrophic denitrification activity by comparing differences in volumetric rates of

1. nitrite-nitrogen consumption between Reactors 4 and 2;
2. ammonium-nitrogen production (through cell decay) between Reactors 4 and 3; and
3. total inorganic nitrogen (TIN) consumption between Reactors 4 and 3.

Under anoxic conditions, heterotrophic denitrifiers consume nitrate, nitrite, and organic carbon; they also produce ammonium via cell decay when other organic carbon is limited (as in this assay). Novel autotrophic denitrifiers consume ammonium and nitrite and produce nitrogen gas. Reactors 4 and 2 both contain nitrite, but with and without ammonium, respectively, so that the effect of ammonium (expected to increase activity of novel organisms) on the nitrite reduction rate between Reactor 4 and 2 can be examined. Reactors 4 and 3 both contain ammonium, but with/without nitrite. Reactor 4 (with nitrite present) would be expected to consume more ammonium if autotrophic denitrifiers are active. With cell decay releasing ammonium, this would then be measured as a decrease in the ammonium-nitrogen production rate in Reactor 4 compared to Reactor 3. Comparing TIN consumption rates, Reactor 4 (with both ammonium and nitrite present) is expected to consume more TIN than Reactor 3 if autotrophic denitrification occurs.

Heterotrophic nitrification activity was determined as the difference in the volumetric ammonium-nitrogen reduction rates and the volumetric nitrate-nitrogen production rates between aerobic batch Reactors 8 and 7. Autotrophic nitrifiers consume ammonium and produce nitrate under aerobic conditions, while heterotrophic nitrifiers consume ammonium and organic carbon and produce nitrate. Other heterotrophs growing on acetate also need ammonium for cell synthesis. Although the proportions of ammonium used for cell synthesis and nitrification are not known, nitrate will be produced from the amount nitrified. Thus, by comparing the nitrate-nitrogen production rate with or without added acetate, it is possible to measure the portion of ammonium disappearance caused by heterotrophic nitrification..

By linear regression of the analytical data versus time using the least squares method, the researchers used Microsoft Excel to calculate zero-order reaction rates and their standard deviation for each of the batch reactors.. The R^2 for fit of the data to the linear assumption typically ranged from 85-95%, with most over 90%, indicating excellent fits. No significant lack of fit was indicated by the residuals. The rates were then compared and tested to determine whether differences were statistically significant using the Student t–test (Draper and Smith, 1966; Ostle, 1963). Standard deviations for the zero-order reaction rates, calculated as above, were pooled for use in completing these analyses.

2.3.3.2 Anaerobic/Aerobic Batch Assay for Phosphorus Removal

The anaerobic/aerobic batch test for PAO activity provided 75 minutes of anaerobic conditions followed by a 3.5-4 hour aerobic period. The same apparatus was used as for the nitrogen removal batch assay. Fresh mixed liquor was collected from the third channel of the Orbal™ oxidation ditch, and 1 L was added to duplicate reactors. A working volume of 1 L was used. DO and pH were monitored, and 30 mg/L as COD of potassium acetate was added to the reactors after purging with N_2 gas to ensure anaerobic conditions. Samples were collected at 15-minute intervals for 75 minutes for the anaerobic period, and at 15 minutes initially and at half-hour intervals after 30 minutes for the aerobic period. After centrifuging, the supernatants were analyzed for phosphate during the anaerobic and aerobic tests, and for nitrate during the anaerobic period.

2.3.4 Results and Discussion

2.3.4.1 Anoxic Batch Reactor Assays

Figure 2-3 provides results from an anoxic batch test using Elmwood mixed liquor. The results are typical of all the anoxic batch tests. Close examination reveals several important trends.

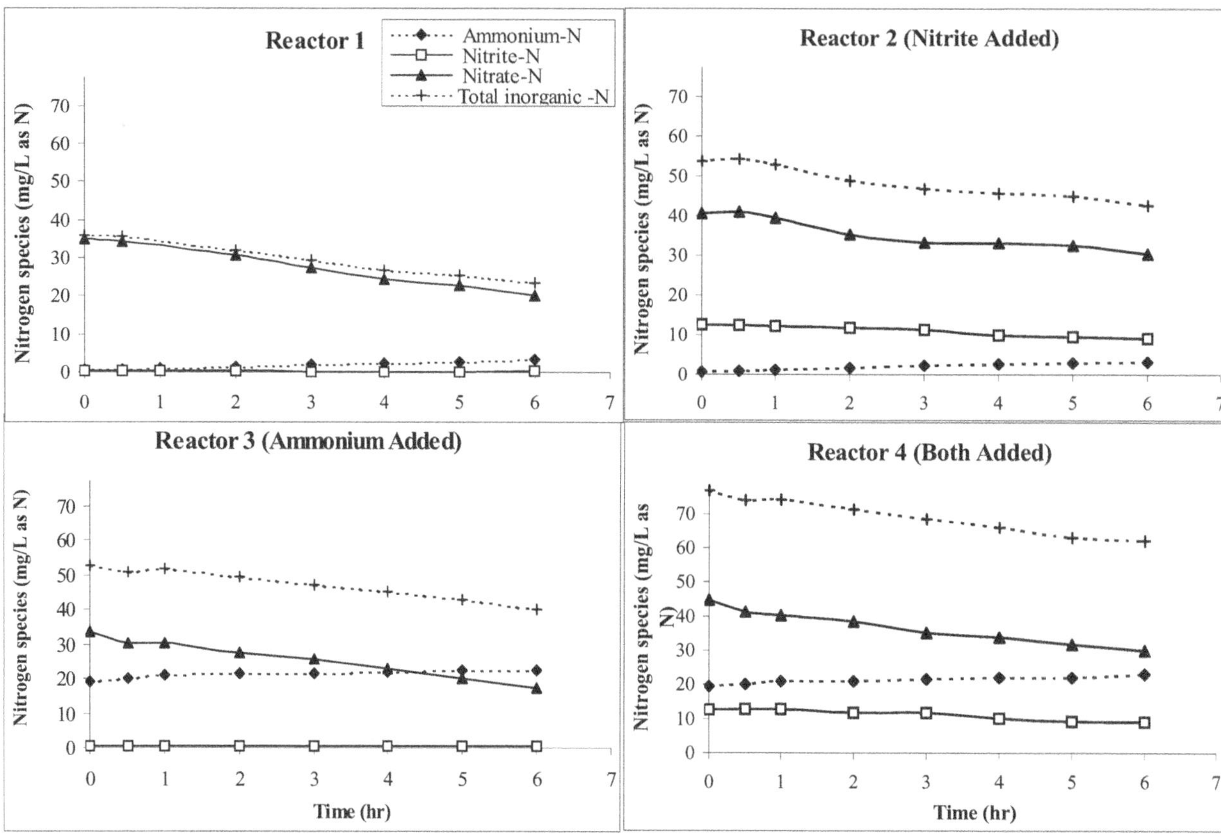

Figure 2-3. Typical Anoxic Batch Assay Result for Elmwood. (Reprinted with permission from Littleton et al., 2003a.)

First, the ammonium concentration increases in all four anoxic batch test reactors. Ammonium release is expected as a result of ammonification of organic nitrogen during cell decay. Since ammonium does not decrease in any of the batch reactors, ammonia consumption through autotrophic denitrification is obviously not substantial in relation to this ammonium release mechanism..

Second, nitrate was consumed in all four reactors, presumably as a result of endogenous respiration by heterotrophs using nitrate as a terminal electron acceptor. This reaction could cause substantial total inorganic nitrogen (TIN) consumption in the batch reactors via conversion to nitrogen gas.

Third, the nitrite concentration did not appear to change in Reactors 1 and 3, where none was added (i.e., no net nitrite was produced in these bioreactors). There was, however, a slight decrease in Reactors 2 and 4 where nitrite was added.

Finally, TIN was consumed in all four reactors. The TIN reduction closely reflects the decrease in nitrate-nitrogen, reduced slightly by the small increase in ammonium-nitrogen. Nitrate was present throughout the entire test for heterotrophic denitrification.

Figure 2-4 provides an expanded view of the ammonium-nitrogen in Reactors 3 and 4 and nitrite-nitrogen in Reactors 2 and 4 for the results provided in Figure 3.2. This illustrates the similarity of the reaction rates whether one or both compounds were added.

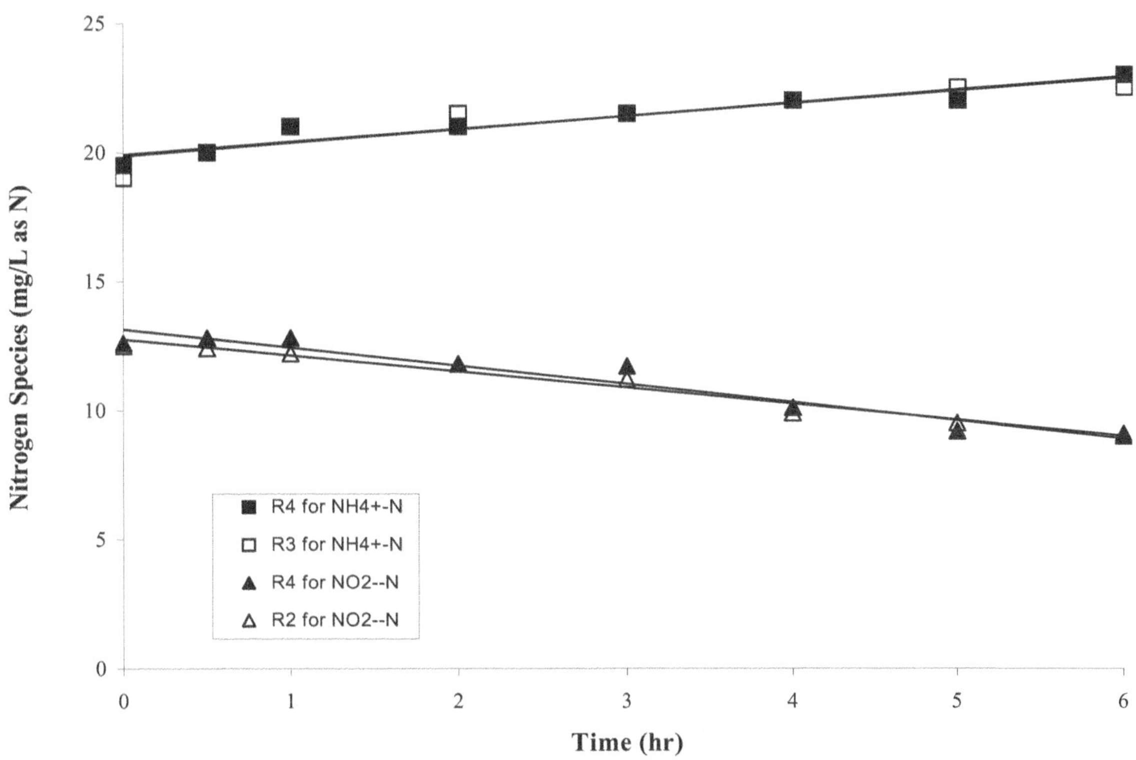

Figure 2-4. Elmwood Anoxic Test for NH_4^+-N and NO_2^--N Reaction Rates. (Reprinted with permission from Littleton et al., 2003a.)

The upper portion of Table 2-7 summarizes the results, including statistical analyses, from the anoxic batch reactor tests for all three plants. The results indicate that, except for Hartford Road, the difference in the nitrite-nitrogen reduction rate between Reactors 2 and 4 was not statistically significant for any of the experimental runs. The rate there was so small that it is probably a statistical anomaly, and in any case was negative (the opposite of the effect expected for autotrophic denitrification). Thus, this comparison did not indicate any autotrophic denitrification activity. Also, the ammonium-nitrogen production rate was not significantly different, except for Hammonton, where it was higher in Reactor 4 (not lower, as would result from autotrophic denitrification). The differences in TIN reduction were statistically significant at Elmwood and Hartford Road, but not at Hammonton. The differences are relatively minor (increased by about 20% when nitrite was added). The comparison is to a denitrification rate (in Reactor 3) where no external carbon source was added to accelerate heterotrophic denitrification, meaning that the heterotrophic denitrification rate is based only on endogenous respiration. In the full-scale plant, exogenous carbon would be present to allow an even higher heterotrophic denitrification rate, further reducing the importance of any novel reactions. It should further be noted that heterotrophic denitrification of nitrite is a three-electron transfer, while that of nitrate is five electrons. This means that 5/3's **(AB suggests: five-thirds)**as much nitrogen is expected to be lost from the system if nitrite is the electron acceptor. Thus, the small decreases in nitrite observed in Reactor 4 can account for the differences in TIN disappearance rates between the two reactors (no nitrite was present in Reactor 3).

Table 2-7. Batch Assays: Volumetric Nitrogen Conversion Rates, Mean ± Standard Error, mg-N/(L-hr).

	Elmwood		Hammonton		Hartford Road
	Run 1	Run 2	Run 1	Run 2	
Anoxic Batch Assay					
NO_2^--N Reduction Rate					
Reactor 4	0.704 ± 0.074		0.20 ± 0.05		0.04 ± 0.024
Reactor 2	0.622 ± 0.045		0.15 ± 0.02		0.13 ± 0.045
Difference	0.082		0.05		-0.09*
NH_4^+-N Production Rate					
Reactor 4	0.496 ± 0.06		0.330 ± 0.036		0.77 ± 0.08
Reactor 3	0.512 ± 0.090		0.24 ± 0.053		0.77 ± 0.06
Difference	-0.016		0.09 *		0
Total Inorganic Nitrogen Reduction Rate					
Reactor 4	2.49 ± 0.12		1.30 ± 0.15		2.18 ± 0.07
Reactor 3	2.04 ± 0.13		1.16 ± 0.09		1.80 ± 0.21
Difference	0.45 *		0.16		0.38 *
Aerobic Batch Assay					
NH_4^+-N Reduction Rate					
Reactor 8	4.31 ± 0.47	5.37 ± 0.24	2.19 ± 0.25	3.07 ± 0.08	4.88 ± 0.34
Reactor 7	4.52 ± 0.62	3.52 ± 0.48	2.07 ± 0.06	2.4 ± 0.10	4.03 ± 0.32
Difference	-0.21	1.85 *	0.12	0.67	0.85 *
NO_3^--N Production Rate					
Reactor 8	4.36 ± 0.47	4.07 ± 0.13	2.17 ± 0.11		
Reactor 7	4.45 ± 0.11	4.06 ± 0.12	2.24 ± 0.08		
Difference	-0.09	0.01	-0.07		

* Difference is statistically significant ($P < 0.05$).

Much of the decrease in total inorganic nitrogen in the anoxic batch assays could have been due to heterotrophic decay under anoxic conditions. This is essentially digestion of the biomass using nitrite or nitrate as the electron acceptor. The stoichiometric relationship between biomass destruction, ammonia release, and nitrate consumption is generally known. Typically, the nitrogen content of biomass is 0.08 to 0.1 mg N/mg biomass COD (the default value for the International Water Association Activated Sludge Models No. 1 and 2 is 0.087 - Grady et al., 1999). Thus, the release of 1 mg/L of ammonia-N is associated with the oxidation of between

(1/0.08 =) 12.5 and (1/0.1 =) 10 mg/L of biomass COD. Since the oxygen equivalent of nitrate is 2.86 mg of O_2 equivalent/mg nitrate-nitrogen (Grady et al., 1999), the nitrate-nitrogen required to oxidize between 10 and 12.5 mg/L of biomass COD is (10/2.86 =) 3.5 to (12.5/2.86 =) 4.4 mg/L as N. Therefore, in the anoxic tests, the ratio of nitrate-nitrogen denitrified to ammonia-nitrogen released through decay is expected to be between about 3.5 and 4.4. Table 2-8 was generated by plotting the nitrate-nitrogen reduced versus the ammonia-nitrogen produced and calculating the relationship between the two using linear regression. The close fit of the experimental results to the theoretical values suggests that biomass decay was the principal mechanism of ammonia production and nitrate removal in the anoxic batch tests.

Table 2-8. Anoxic Batch Assay–Ratio of NO_3^--N Denitrified to NH_4^+-N Released and First Order Decay Rate[1].

	Elmwood Run 1	Hammonton Run 1	Hartford Road
NO_3^--N Denitrified to NH_4^+-N Released (mg NO_3^--N / mg NH_4^+-N) Mean Value of Ratio ± Standard Error			
Reactor 1	----	4.93 ± 0.53	4.71 ± 0.22
Reactor 2	4.07 ± 0.28	4.56 ± 0.75	4.00 ± 0.17
Reactor 3	4.16 ± 0.75	4.41 ± 0.90	3.88 ± 0.19
Reactor 4	4.32 ± 0.45	4.49 ± 1.42	3.60 ± 0.28
Average	4.18 ± 0.52	4.49 ± 1.02	3.83 ± 0.22
Expected Value[2]	3.50 - 4.37	3.50 - 4.37	3.50 - 4.37
First Order Decay Rate (d^{-1})			
Reactor 1	0.037	0.046	0.046
Reactor 2	0.026	0.035	0.051
Reactor 3	0.036	0.031	0.048
Reactor 4	0.043	0.036	0.047

[1] Typical first order decay rate for activated sludge is 0.05 d^{-1}.
2 Calculated for biomass nitrogen content range of 0.08 - 0.1 mg N/mg COD.

The researchers extended the analysis by considering the kinetics of denitrification. If biomass decay is the principal mechanism, then the nitrate-nitrogen destruction rate can be used to estimate the first order specific decay rate using nitrate as a terminal electron acceptor. The volumetric nitrate-nitrogen destruction rate is determined by linear regression of the nitrate-nitrogen concentration versus time, with the result expressed in units of mg nitrate-N/L-hr. Multiplying by 2.86 mg O_2 equivalent/mg nitrate-N gives the COD destruction rate in mg COD/L-hr. Dividing by 1.42 mg COD/mg VSS (typical for biomass, Grady et al., 1999) gives the VSS destruction rate in mg VSS/L-hr. Dividing this by the MLVSS concentration in the test reactor gives the first order specific decay rate in units of hr^{-1} (mg VSS/mg VSS-hr), which can subsequently be converted to units of day^{-1}. The calculated values, presented in Table 2-8, are in the range of 0.03 to 0.05 day^{-1}. The specific decay rate under anoxic conditions should be similar to, but lower than, the specific decay rate using molecular dissolved oxygen as the terminal electron acceptor (Grady et al, 1999). The two specific rates are related by the η factor, which relates anoxic and aerobic activity in single sludge denitrification systems (Henze et al., 1999). Typical aerobic first order specific decay rates for activated sludge are about 0.05 day^{-1} (Metcalf & Eddy, 1991), which is similar to or somewhat greater than the first order specific decay rates under anoxic conditions reported in Table 2-8. This further suggests that heterotrophic decay can account for the loss of nitrogen observed in the anoxic batch reactors.

Taken together, the results of the anoxic batch assays suggest only a very limited role, at most, for autotrophic denitrification in nitrogen removal at the three plants investigated. This conclusion is supported by the following observations.

1. There was no statistically significant difference in nitrite reduction between Reactors 4 (with ammonia added) and 2 (with no ammonia added), except for Hartford Road, where the rate in Reactor 2 was actually greater (when it should have been smaller if autotrophic denitrification were occurring).
2. There was no statistically significant difference in ammonia production rates between Reactors 4 (with nitrite added) and 3 (with no nitrite added), except for Hammonton, where the rate in Reactor 4 was actually greater (when it should have been smaller if autotrophic denitrification were occurring).
3. The total inorganic nitrogen reduction rate increase in Reactor 4 (with nitrate added) was slightly greater than in Reactor 3 (with no nitrite added), but that can be explained by the fact that denitrification with nitrite as the terminal electron acceptor is a 3 electron transfer while that with nitrate is a 5 electron transfer.
4. The stoichiometry of ammonia production and nitrate reduction in the batch reactors was consistent with the decay of heterotrophic biomass using nitrate as the terminal electron acceptor.
5. The kinetics of nitrate reduction in the batch reactors were consistent with the decay of heterotrophic biomass using nitrate as the terminal electron acceptor

2.3.4.2 Aerobic Batch Reactor Assays

Figure 2-5 provides results for a typical aerobic batch test—this one from the Hammonton plant. Close examination reveals several important trends.

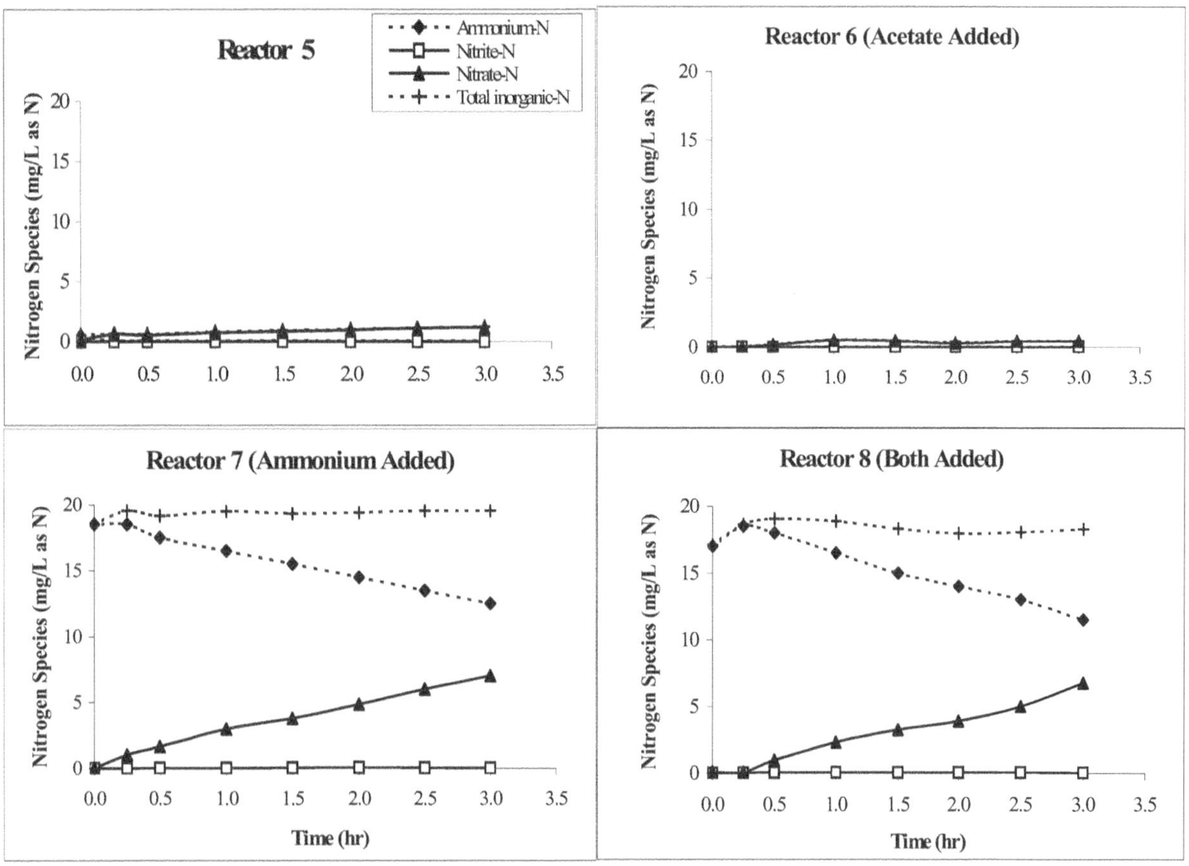

Figure 2-5. Typical Aerobic Batch Assay Results for Hammonton. (Reprinted with permission from Littleton et al., 2003a.)

First, very little inorganic nitrogen is present in either Reactors 5 or 6. Some inorganic nitrogen production is suggested in Reactor 5, where no ammonia or acetic acid was added. This is probably a result of biomass decay, resulting in the release of ammonia-nitrogen, which is subsequently nitrified. The growth of heterotrophs as a result of acetic acid addition to Reactor 6 probably consumed the ammonia released by biomass decay, resulting in essentially no change in inorganic nitrogen.

Second, substantial nitrification occurs in both Reactors 7 and 8, as evidenced by the decline in ammonia and the corresponding increase in nitrate. Note that little nitrite accumulates in either of these reactors, suggesting that ammonia oxidation to produce nitrite, and nitrite oxidation to produce nitrate, are balanced.

Finally, the TIN concentration remains essentially constant in both Reactors 7 and 8. Thus, there is essentially no net production or consumption of inorganic nitrogen in these two reactors.

Figure 2-6 compares the ammonia consumption and nitrate production in Reactors 7 and 8. The results demonstrate the essentially stoichiometric conversion of ammonia-nitrogen to nitrate-nitrogen.

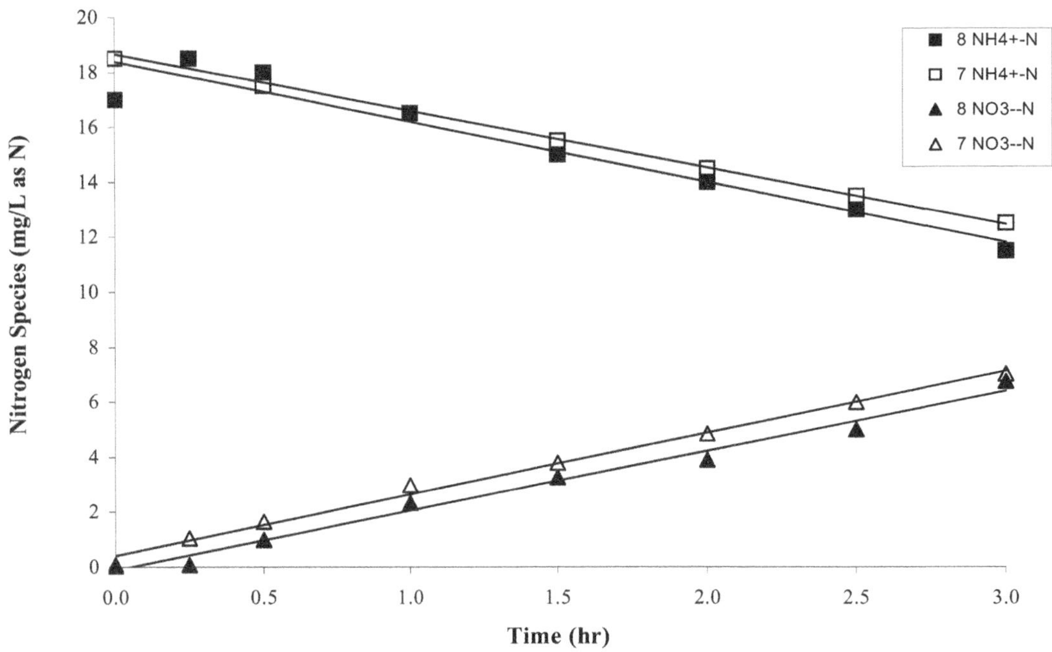

Figure 2-6. Hammonton Aerobic Test for NH_4^+-N and NO^{3-}-N Reaction Rates. (Reprinted with permission from Littleton et al., 2003a.)

Table 2-7 (lower portion) summarizes the results from the aerobic batch reactor tests from all three plants. The results show that the difference in ammonia-nitrogen reduction rates between Reactors 7 and 8 is statistically significant in Elmwood run 2 and Hartford Road, but not in the other three cases. However, the nitrate-nitrogen production rates are not significantly different between Reactors 7 and 8 for any of the tests. Since many heterotrophs can grow on acetate and use ammonia-nitrogen for cell synthesis, the higher ammonia-nitrogen consumption rate in Reactor 8, where acetate was added, is probably due to higher cell synthesis. The nitrate-nitrogen production rate shows no difference with or without acetate. This indicates that, under aerobic conditions, autotrophic nitrifiers, not heterotrophic nitrifiers, carry out essentially all of the observed nitrification.

No nitrite build-up was observed in any of the aerobic batch tests. This suggests that the original mixed liquor contained balanced populations of ammonia and nitrite oxidizers. This is important because, if substantial autotrophic denitrification were occurring, nitrogen species would convert from ammonia to nitrite to nitrogen gas. This would reduce the amount of nitrite available for oxidation and hence decrease the number of nitrite oxidizers. However, the apparent balance of the populations in the aerobic tests suggests that autotrophic denitrification was not directing a substantial amount of nitrite-nitrogen to nitrogen gas before nitrite oxidizers could utilize it. The apparent balance also suggests that, in the full-scale plant, ammonia was oxidized fully to nitrate (rather than just to nitrite) before it was subsequently reduced to nitrogen gas by denitrification. If this were not the case, then the populations of ammonia and nitrite oxidizers would not be in balance.

The conclusion that autotrophic denitrification and heterotrophic nitrification are not substantial is consistent with the results of others. Those results include the full-scale operating data and modeling analysis conducted by Daigger and Littleton (2000), and the literature review by Van Loosdrecht and Jetten (1998).

2.3.4.3 Anaerobic/Aerobic Batch Assays for Phosphorus Removal

Figure 2-7 shows results from two anaerobic/aerobic batch assays from the Elmwood plant. Rapid and substantial phosphate release occurred during the anaerobic period in response to the acetate added to begin this period. Approximately 15 mg/L as P of dissolved phosphate is released as a result of the addition of 30 mg/L of acetate. Typically, 0.4 to 0.6 mg as P of phosphate is released by PAOs for every mg of acetate taken up in the anaerobic zone (IWA, 2000; Henze et al., 1999; Filipe and Daigger, 1998; Barker and Dold, 1997a and b). Thus, the magnitude of the observed phosphate release is consistent with the observed stoichiometry of PAOs. It is also interesting to observe that substantial phosphate release is obviously occurring, even though such release is not typically observed in the full-scale facility (Daigger and Littleton, 2000).

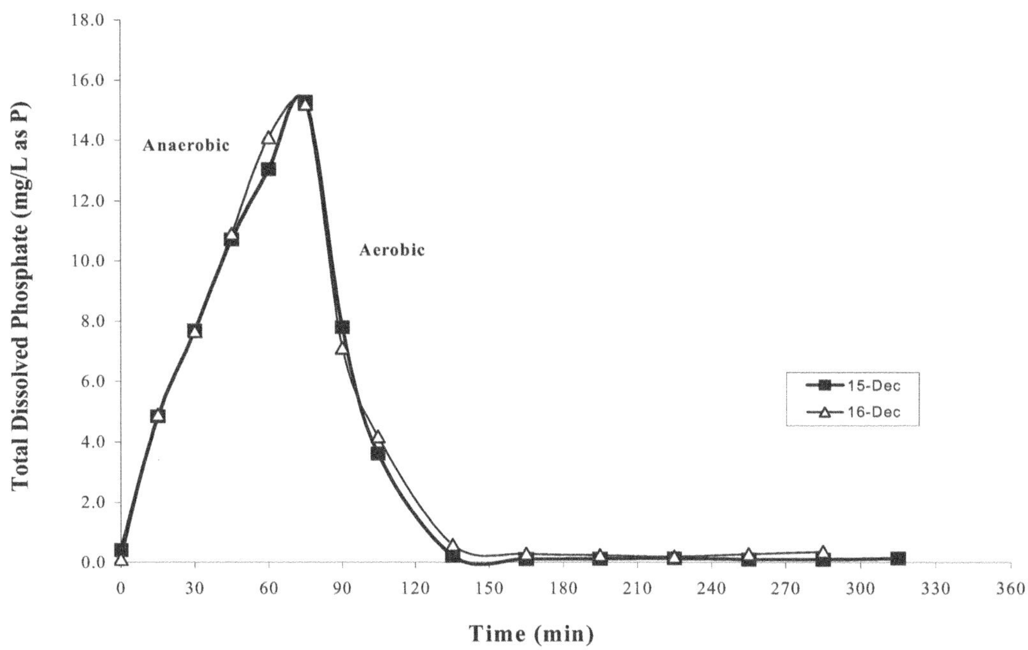

Figure 2-7. Elmwood Sludge Phosphorus Metabolism during Anaerobic/Aerobic Batch Assay. (Reprinted with permission from Littleton et al., 2003a.)

Rapid uptake of anaerobically released phosphate began with the onset of aerobic conditions, and was nearly complete within about 60 minutes. The measured zero-order phosphate uptake rate is 562 mg P/L-d. According to ASM2d, the specific uptake rate for PAOs is 1.5 g P/g of VSS PAOs-day (IWA, 2000; Henze et al., 1999). Based on the observed bulk phosphate uptake rate and this specific rate, the PAO concentration can be calculated to be 374 mg/L as VSS, which is 12.6% of the total mixed liquor (MLVSS = 2960 mg/l).

The anaerobic/aerobic batch assay clearly demonstrates a pattern of anaerobic phosphate release (following acetate addition) and aerobic phosphate uptake, as expected for PAOs. The stoichiometry of phosphate release was as expected, and phosphate uptake was rapid. Based on the volumetric phosphate uptake rate, it was estimated that approximately 13% of the process mixed liquor consisted of PAOs. Interestingly, Noguera et al. (2000) and Huang et al. (2002) have reported PAOs to account for ~14% of the biomass at two Orbal^{TM} plants.

Previous observations at the Elmwood plant had indicated that: 1) more phosphate was being removed than was expected simply for biomass synthesis , 2) the phosphorus content of the mixed liquor was elevated (2.5% P/VSS), and 3) bacteria containing polyphosphate granules were observed microscopically within the biomass (Daigger and Littleton, 2000). The current anaerobic/aerobic batch assay results, coupled with these previous observations, strongly suggest that biological phosphorus removal is occurring in those plants and that the characteristics of the organisms conform with our current understanding of the characteristics and behavior of PAOs. These results are consistent with recent research that demonstrates the presence of PAOs in other Orbal™ systems (Zilles et al., 2001, 2002). Interestingly, despite the substantial phosphate release observed with the process mixed liquor in the batch test (Figure 2-7), such release is not observed in the full-scale facility (Daigger and Littleton, 2000). This may be a result of a mixing pattern developing in the full-scale facility that provides multiple occasions of phosphate release and uptake, in small increments, resulting in little observed increase in dissolved phosphate. In contrast, other observers (Zilles et al., 2002). have reported phosphate release (up to 12 mg/L) in the outer channel of full-scale plants. Presumably, this discrepancy stems from differences in operating conditions—perhaps including creation of a fully anaerobic environment in the outer channel that resulted in phosphate release.

2.3.4.4 Analysis of Batch Test Methodology

One of the objectives of this research was to develop simple and effective assay methods to determine whether the activities of autotrophic denitrifiers and/or heterotrophic nitrifiers are significant in full-scale SBNR systems. The batch assay screening techniques used provided a method for assaying these activities. The results also indicated that, for the three plants investigated, the activities of these organisms were not significant.

The results also suggest that the subject test procedure can be refined to be more effective and efficient. For the anoxic batch tests, the differences in ammonia production rates between Reactors 4 (with ammonia and nitrite added) and 3 (with only ammonia added) were the most useful in determining whether significant autotrophic denitrification was occurring. Thus, rather than operating Reactors 1 and 2, duplicate batch reactors could be operated to improve the statistical significance of the results. Alternatively, Reactor 2 (with nitrite but no ammonia added) could be operated to allow nitrite reduction rates to be compared to Reactors 4 (with ammonia and nitrite added). Reactor 1 (with no ammonia or nitrite added) provided little additional information to discriminate the relative contributions of autotrophic and heterotrophic denitrifiers. Direct comparison of the total inorganic nitrogen reduction rates between Reactors 4 (with ammonia and nitrite added) and 3 (with only ammonia added) was complicated by the fact that nitrite served as a terminal electron acceptor for heterotrophs in Reactor 4.

The results of the aerobic batch tests indicated that only Reactors 7 (with ammonia added) and 8 (with ammonia and acetate added) were needed to assay aerobic heterotrophic nitrification activity. The activity in Reactors 5 and 6, with no ammonia added, was minimal and was not expected to be substantial for either autotrophic or heterotrophic nitrifiers. These reactors were operated in this research for completeness but contributed little to understanding the conversions occurring in the batch reactors. An alternative approach would be to operate four reactors but to duplicate the conditions provided in Reactors 7 and 8 to improve the power of the statistical analysis. The researchers observed that, due to the well-known consumption of alkalinity by nitrification, pH control in the aerobic batch assay required consistent attention. The analyst must be prepared to provide the necessary attention or, alternatively, an automatic pH control system could be provided.

2.3.4.4 Observation of SBNR in Batch Reactors

Errors in the initial set-up of the anoxic batch tests actually gave rise to some insight into the conditions that can lead to SBNR in these systems. The researchers found that SBNR occurred using mixed liquor from the full-scale plant when inadvertent aeration was occurring even at so low a rate that a bulk DO concentration was not detected. Under these conditions, sufficient oxygen apparently was being provided to allow nitrification to occur, but not to produce a residual DO concentration that would prevent denitrification.

In those initial anoxic batch assays, the three-way needle valve used to allow gas in and out of the laboratory reactors was not sealed, and substantial ammonia and nitrite reduction occurred (Figure 2-8). Nitrate accumulation was not observed (results not shown). The ammonia-nitrogen reduction rate was nearly half the rate observed in the aerobic test in Reactors 7 or 8, even though no intentional aeration was provided. One might conclude that the nitrification rate in the test depicted in Figure 2-8 was limited by the oxygen transfer rate. Denitrification was also observed as the added nitrite was quickly reduced. Although nitrogen gas was purged through the reactor before the reaction began, and the DO meter showed a value of zero, it appeared that oxygen was still transferred into the system, presumably due to the high DO deficit and the large DO demand in the system. This oxygen transfer allowed nitrification to occur at DO concentrations that still allowed added nitrite and any nitrite/nitrate formed by nitrification to be denitrified. This suggests that the nitrifiers present in the full-scale mixed liquor have a high affinity for dissolved oxygen. This observation is consistent with recent results obtained from the study of another Orbal™ system (Park et al., 2001, 2002).

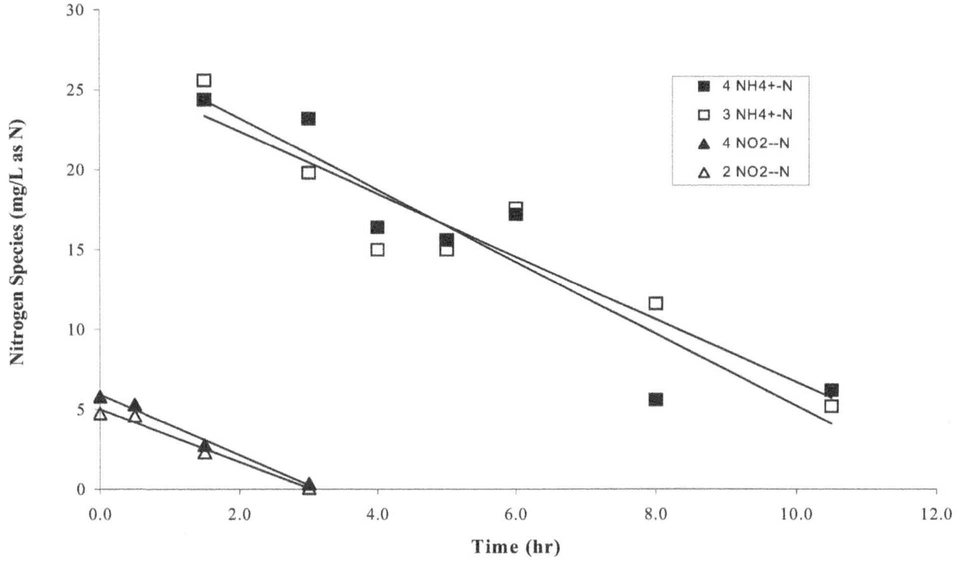

Figure 2-8. Elmwood Anoxic Batch Assay with Air Leak. (Reprinted with permission from Littleton et al., 2003a.)

Parallel clean water tests with and without the valve open were subsequently conducted to determine whether anoxic conditions were clearly established. Using nitrogen gas, it took more than 20 minutes to purge the tap water to zero DO, while it only took two minutes to purge mixed liquor. When mixing of tap water was continued without aeration and without nitrogen purging, the DO concentration in a reactor with an open needle valve increased to nearly 3 mg/L in 40 minutes. In contrast, a reactor with a closed valve during the test (the technique subsequently used in the anoxic test) was able to maintain zero DO. Thus, although the DO meter showed a zero reading during the preliminary anoxic test, oxygen was still being transferred via the open valve. Further testing with mixed liquor showed that, with the valve closed, ammonium-nitrogen did not decrease, demonstrating that anoxic conditions were created. These results indicate that at low DO, oxygen was transferred efficiently even without aeration. The results also indicate that at low DO, nitrification and denitrification can occur simultaneously in the mixed liquor from the full-scale system tested. Finally, with regard to the test procedure, these results highlight the importance of demonstrating that no oxygen transfer is occurring in the laboratory reactor used for the batch tests. Substantial oxygen transfer can occur even when the residual DO concentration is negligible and the circumstances suggest that little oxygen transfer would be expected.

2.3.5 Summary and Conclusions

Based on the known biochemistry of the microorganisms considered, the researchers developed a series of batch assays as a screening tool to identify the activity of autotrophic denitrification under anoxic conditions and heterotrophic nitrification under aerobic conditions in bioreactors accomplishing SBNR. Autotrophic denitrification was tested under anoxic conditions using a 2 by 2 experimental design with ammonium and/or nitrite addition. Heterotrophic nitrification was tested under aerobic conditions using a 2 by 2 design with ammonium and acetate addition. The test procedure was applied to three full-scale plants using the Orbal™ process, plants that had previously been shown to exhibit significant SBNR. Based on the results they obtained, the researchers suggest that the test procedure for the aerobic batch assay be simplified to include only two batch reactors, with ammonia added to both reactors and acetate added to only one.

Results from the anoxic batch assay demonstrated that ammonia did not decrease under anoxic conditions but, in fact, increased. The principal cause of nitrate consumption and ammonia-nitrogen production was heterotrophic biomass decay. Stoichiometric calculations and denitrification kinetics agree well with the theoretically expected values and support this conclusion. Under aerobic conditions, the differences in ammonia-nitrogen reduction rates between Reactors 7 (no acetate) and 8 (acetate added) were small and probably the result of increased assimilation for growth. The differences in nitrate-nitrogen production between the two reactors were not statistically significant. Thus nitrification appeared to result from the activity of autotrophic nitrifiers, not heterotrophic nitrifiers (whose activity would be expected to increase with acetate addition). Batch test results with a very low aeration rate also demonstrated that nitrification and denitrification could occur simultaneously without detected DO.

Anaerobic/aerobic batch assays assessed the activity of PAOs at one plant. Phosphate release was consistent with observed stoichiometry, and the zero-order phosphate uptake rate would suggest that PAOs comprised approximately 13% of the process mixed liquor.

The following conclusions were reached.

1. The simple and effective screening tool developed enables autotrophic denitrification and heterotrophic nitrification activity to be assayed using mixed liquor from full-scale plants.
2. If they occurred at all, the rates of autotrophic denitrification and heterotrophic nitrification were very low, and did not play a major role in nitrogen transformations for these systems. The results indicate that novel microorganisms are not substantial contributors to SBNR in the staged, -loop bioreactors studied.
3. Balanced populations of ammonia- and nitrite-oxidizing bacteria were present in the full-scale mixed liquors tested, as evidenced by the lack of nitrite accumulation during the aerobic batch assays. This suggests that ammonia is being converted to nitrate in the full-scale systems before it is subsequently denitrified.
4. Following acetate addition, anaerobic phosphate release and aerobic phosphate uptake were observed in the anaerobic/aerobic batch assay, as expected based on current understanding of PAO metabolism. These observations, coupled with previous observations of process phosphate removal, mixed liquor phosphate content, and the presence of intracellular polyphosphate granules, strongly suggest that excess biological phosphorus removal was occurring in these systems. These results also demonstrate that the mixed liquor was capable of phosphate release, even though such release was not observed in the full-scale system.

CHAPTER 3.0

COMPUTATIONAL FLUID DYNAMICS

3.1 Background

Computational fluid dynamics (CFD) is the science of determining a mathematical solution to fluid flow through space and time. It entails the analysis of systems involving fluid flow, heat transfer, and associated phenomena such as chemical reactions, by means of computer-based simulation. The technique is very powerful and spans a wide range of industrial and nonindustrial applications. Generally, the water and wastewater industries have been more concerned with the chemical compounds present than with the fluid in which they react. Although CFD has been used to analyze flow distributions in storage, disinfection contact, and settling tanks (Ducoste and Daigger, 1999), and in pumps and mixing/injecting systems, little effort has been focused on activated sludge biochemical reactions. Until now, there has been no report on a direct examination of activated sludge biochemical reactions combined with the fluid mixing pattern in a CFD environment.

It is generally recognized that spatially varying environments can develop within bioreactors with high mixed liquor recirculation rates, such as in closed loop bioreactors, and basins that are aerated with surface mechanical aerators (Cinar et al., 1998; Randall et al., 1992). It is often assumed, though, that flow in closed loop bioreactors exhibits a laminar pattern. However, preliminary velocity measurements and visual observations in a closed loop bioreactor test tank (Envirex, 1999) suggested that a rotational spiral-type flow coexists with the main recirculating flow. This "second current" is actually a well-known phenomenon in curved pipe or tank environments (Wei et al., 1992).

It is hypothesized here that this second current in closed-loop bioreactors, along with the main mixed liquor recirculation flow, could create alternating anaerobic and aerobic conditions, both vertically and horizontally, which would allow SBNR to occur. So, for reliable nutrient removal, it may be crucial to gain an understanding of these mixing patterns, and CFD could detectthose zones. Using this understanding and a tool such as CFD, it might be possible to make small, inexpensive changes in tank flow pattern that could substantially improve SBNR reliability and performance.

Visualization has always played an important role in understanding flow phenomena and in physical-intuitive reasoning. The unprecedented increase of computer power, allied with development of efficient visualization tools, has considerably and irreversibly changed the landscape of fluid dynamics in both its fundamental and its applied aspects. Precise and fast numerical algorithms now allow one, in some cases, to solve deterministically the equation of fluid motion, in particular for phenomena that are rapidly changing in time and space, such as turbulence, vortices, or shock waves. A new concept of "numerical experiments" has appeared, complementary to laboratory experiments, and sometimes replacing them (Fletcher, 1991).

The general objective of this research was to characterize the flow and mixing pattern in a full-scale closed loop bioreactor and examine their effect on SBNR. This chapter describes the basic CFD model and software used.

3.2 Computational Fluid Dynamics Software

The CFD software package CFX4 (ANSYS, formerly AEA Technology, Inc, Pittsburgh, PA) was used to model fluid flow in this research. CFD models typically include a description of the flow geometry, a set of coupled differential equations describing the physics and chemistry of the flow, boundary and initial conditions, and a structured mesh of points at which these equations are solved (Warsi, 1991).

3.3 Governing Equations

The equations of motion in CFX4 are solved by a finite volume technique (Lesieur et al., 1996). The software is based on the fundamental governing equations of fluid dynamics—the continuity, momentum, and energy equations. These governing equations are programmed in CFX4 for conservation of mass and momentum, using the Navier-Stokes equations. Experience shows that the Navier-Stokes equations accurately describe Newtonian flow. Strictly speaking, "Navier-Stokes" refers only to the components of the momentum equation; however, it is common practice to include the continuity and energy equations in the set of equations referred to by this term. There are several forms of the Navier-Stokes equations (Anderson, 1995) depending on how they are derived, but the most useful here (Versteeg and Malalasekera (1995) are shown in Table 3-1.

Table 3-1. Navier-Stokes Equations, Conservative Form of Governing Equations for Newtonian Flow.

Mass	$\frac{\partial \rho}{\partial x} + \nabla(\rho \mathbf{u}) = 0$ or $\frac{\partial \rho}{\partial x} + div\,(\rho \mathbf{u}) = 0$	3.3-1
x-momentum	$\frac{\partial(\rho u)}{\partial t} + div(\rho u \mathbf{u}) = -\frac{\partial p}{\partial x} + div\,(\mu \ grad\, u) + S_{Mx}$	3.3-2
y-momentum	$\frac{\partial(\rho v)}{\partial t} + div(\rho v \mathbf{u}) = -\frac{\partial p}{\partial y} + div(\mu\, grad\, v) + S_{My}$	3.3-3
z-momentum	$\frac{\partial(\rho w)}{\partial t} + div(\rho w \mathbf{u}) = -\frac{\partial p}{\partial z} + div\,(\mu\, grad\, w) + S_{Mz}$	3.3-4
Internal energy	$\frac{\partial(\rho i)}{\partial t} + div(\rho i \mathbf{u}) = -\, p\, div\, \mathbf{u} + div(k\, grad\, T) + \Phi + S_i$	3.3-5

where: vector operator ∇ is defined as $\nabla = \frac{\partial}{\partial x}\mathbf{i} + \frac{\partial}{\partial y}\mathbf{j} + \frac{\partial}{\partial z}\mathbf{k}$; u, v, w = components of the fluid velocity vector **u** in x, y, z direction; divergence velocity $\nabla \mathbf{u} = div\,\mathbf{u} = \frac{\partial u}{\partial x} + \frac{\partial v}{\partial y} + \frac{\partial w}{\partial z}$ (mathematical meaning; its physical meaning is the rate of change of velocity with time for a control volume of a moving fluid element, per unit volume);
$\mathbf{u}\nabla\phi = \mathbf{u}\,\mathrm{grad}\phi = u\frac{\partial \phi}{\partial x} + v\frac{\partial \phi}{\partial y} + w\frac{\partial \phi}{\partial z}$; ρ = density; μ = viscosity constant; $s_{Mx,y,z}$ = momentum source term in x, y, z direction; p= pressure; Φ = dissipation function; *i* = internal energy of a fluid; S_i = source term for internal energy.

It is clear from the above equations that there are commonalties between the various momentum equations and scalar transport equations (such as equation 3.3-6 below). The flow calculations in the CFX software are based on solution of the Navier-Stokes equations with various extensions such as momentum, heat transfer, or mass fraction of a component of the fluid mixture. The default condition is steady, three-dimensional, incompressible, isothermal, single phase, laminar flow. If a general scalar (non-vector) variable ϕ is introduced in the conservative form of all the fluid flow equations, the equations for scalar quantities such as temperature (heat transfer associated with energy equation) and DO concentration (as mass fraction associated with continuity equation) can be written in a generic equation as in the advection-diffusion equation:

$$\frac{\partial(\rho\phi)}{\partial t} + div\,(\rho\,\mathbf{u}\,\phi) - div\,(\Gamma\,grad\,\phi) = S \qquad 3.3\text{-}6$$

or, using other common notation,

$$\frac{\partial(\rho\phi)}{\partial t} + \nabla \bullet (\rho\,\mathbf{u}\,\phi) - \nabla \bullet (\Gamma\,\nabla\,\phi) = S$$

where S is a source or sink term representing creation or destruction of ϕ, (*e.g.*, ϕ can be the momentum or DO), Γ is the diffusion coefficient, and **u** = (u,v,w) is the fluid velocity vector. Integrating the equation over the control volume *dV* or *ndA* (where *dV* = product of normal unit vector, *n*, and area, *A*) gives

$$\underset{\text{transient}}{\int\frac{\partial(\rho\phi)}{\partial t}dV} + \underset{\text{convection}}{\int\rho\phi\,\mathbf{u}\bullet ndA} - \underset{\text{diffusion}}{\int\Gamma\nabla\phi\bullet ndA} = \underset{\text{source}}{\int SdV} \qquad 3.3\text{-}7$$

In words, the conservation of a fluid component in the control volume can be expressed as:

Rate of change of ϕ in control volume with respect to space & time	+	Net rate of decrease of ϕ due to convection out of control volume	−	Net rate of increase of ϕ due to diffusion into control volume	=	Net rate of creation of ϕ within control volume

This is the key step of the finite volume (or control volume) method. All the terms in the equation are then discretized in time and space, and linear equations are obtained.

When dealing with turbulence, the kinetic coefficient κ and energy constant ε can be introduced using time-averaged Navier-Stokes equations, known as Reynolds-averaged equations, to solve the continuity equation (3.3-1), the momentum equations (3.3-2 through 3.3-4), and the scalar transport equation (3.3-6).

Transport equations for additional scalar quantities, such as biomass and substrate concentrations, may be solved along with the flow equations (default scalar is zero in generic equation 3.3-6). When user scalar equations are being solved, the molecular diffusivity of each species must be specified. If ϕ represents an additional species mass fraction, then $\Gamma = \rho\, D_{\phi}$, where D_{ϕ} has the usual definition of the scalar's diffusivity. When S is a source term, it must be specified as non-zero either using command language for a relatively simple source or using the CFX 4.4 User Subroutine USRSRC. When USRSRC is called, the equations have been linearized and are expressed in the steady state form (Anon., 1997):

$$\text{Convection - Diffusion} = \text{Sources}$$

3.4 Applying the Model

Transient flow can be modeled either with fixed time steps in command language or with an adaptive time-stepping procedure that provides automatic control of the time step so that convergence is achieved, avoiding the need for constant user intervention in long transient runs. Although the CFD simulation used in this research did not run transient flow, an under-relaxation factor (URF) was applied to the steady state as a form of pseudo-time evolution, with a different effective time step in each cell. A small URF corresponds to a short time step approaching infinitesimal; it can lead to a faster solution of the linear equations because it will yield a more diagonally dominant matrix, minimizing the non-linearity.

Since these equations have been obtained by integrating the appropriate conservation law equations over a control volume, source terms have the form of the product of the source term and the control volume. Hence the sources in the transport equations have the dimensions of mass/time (kg/s).

The system geometry was defined as one or two source regions (disc blocs) and the tank region (the rest of tank, excluding the disc blocs). The momentum source and oxygen source were both added in the disc region as a 3-D patch based on the known plant operation.

For simulation of a full-scale tank, the inlet boundary was specified as a 2-D patch (as an influent flow dropping to the tank through an open pipe) over the 3-D tank. The tank walls (side and bottom) were specified as a nonslip boundary condition, which means that velocities at the boundary walls are equal to zero. The top, free surface of the liquid in the tank was specified as a slip wall, subject to the fluid velocity. The outlet was at the bottom of the tank, with a 0.91 m x 0.91 m (36" x 36") square gate located on the inner wall. It was set as a 2-D patch pressout (governed by static pressure), which is often the choice when chemical compounds are involved. In some runs a small flow and low scalar concentration (some species as low as 0.01 mg/l, which is 10^{-8} mass fraction unit, as used in CFX4.4) mixed with a large recirculating flow needed to be resolved, so round-off error could be crucial. Therefore, the researchers applied, double precision where appropriate, despite the longer computer run time required.

Additional FORTRAN subroutines were used in some simulations. USRCVG is called at the end of each iteration to allow users to modify and set their own convergence criteria during the run. USRINT allows overwriting the default initial conditions and writing final calculated quantities, such as volume. USRTRN is called after the initial estimate and at the end of each time step. It can be used to monitor the calculation or to produce special output for each step to examine the progress of the solution. In some simulations it was particularly useful for examining the mass balance at the effluent boundary.

CHAPTER 4.0

USE OF COMPUTATIONAL FLUID DYNAMICS TO CHARACTERIZE MIXING PATTERNS AND DISSOLVED OXYGEN GRADIENTS IN CLOSED-LOOP BIOREACTORS

4.1 Introduction

The objective of the work described in this chapter was to use computational fluid dynamics (CFD) to characterize the flow and mixing pattern in a full-scale closed-loop bioreactor. The researchers then explored potential variations in DO concentrations that may occur within the bioreactor.

Velocity distribution data were available from a closed-loop test tank (Dobbeck, 1999). A CFD model was first developed and calibrated for this tank to refine procedures for hydrodynamic modeling of such systems. The purpose of this testing was to verify the CFD modeling technique under controlled and carefully monitored conditions. Modeling of this system facilitated the development and demonstration of CFD modeling procedures.

A CFD model was then developed for the outer channel of the Elmwood Orbal™ wastewater treatment plant and compared to velocity data collected at that facility. Next, simplified oxygen source and sink terms were added to the CFD model of the full-scale facility to examine potential DO concentration variations. The results were evaluated to determine the reactor microenvironment's potential contribution to SBNR.

4.2 Materials and Methods

4.2.1 Test Tank

4.2.1.1 Configuration

The Envirex test tank was 12.2 m (40') long and 6.1 m (20') wide, with a 2.4 m (8') side water depth. It was configured as a closed-loop bioreactor containing disc aerators of the same type used in the full-scale Orbal™ system to be tested later. There were five discs, each 137 cm (54") in diameter, installed in the test tank across a section of the channel. The discs were 33 cm (13") apart. Velocity data were available for a test with a disc submergence of 38 cm (15") and at disc rotational velocities of 29, 43, 55 and 60 revolutions per minute (rpm). Average velocities across the test cross-section were 0.28, 0.38, 0.43, and 0.45 m/sec, respectively, for the four disc rotational velocities. The measured velocity data were obtained at a relatively stable flow cross section directly opposite the disc aerators (Figure 4-1). Measurements were made 38, 84, and 152 cm (15, 33, and 60") away from the inner wall, each at three depths below the water surface (55.2, 110, and 166 cm).

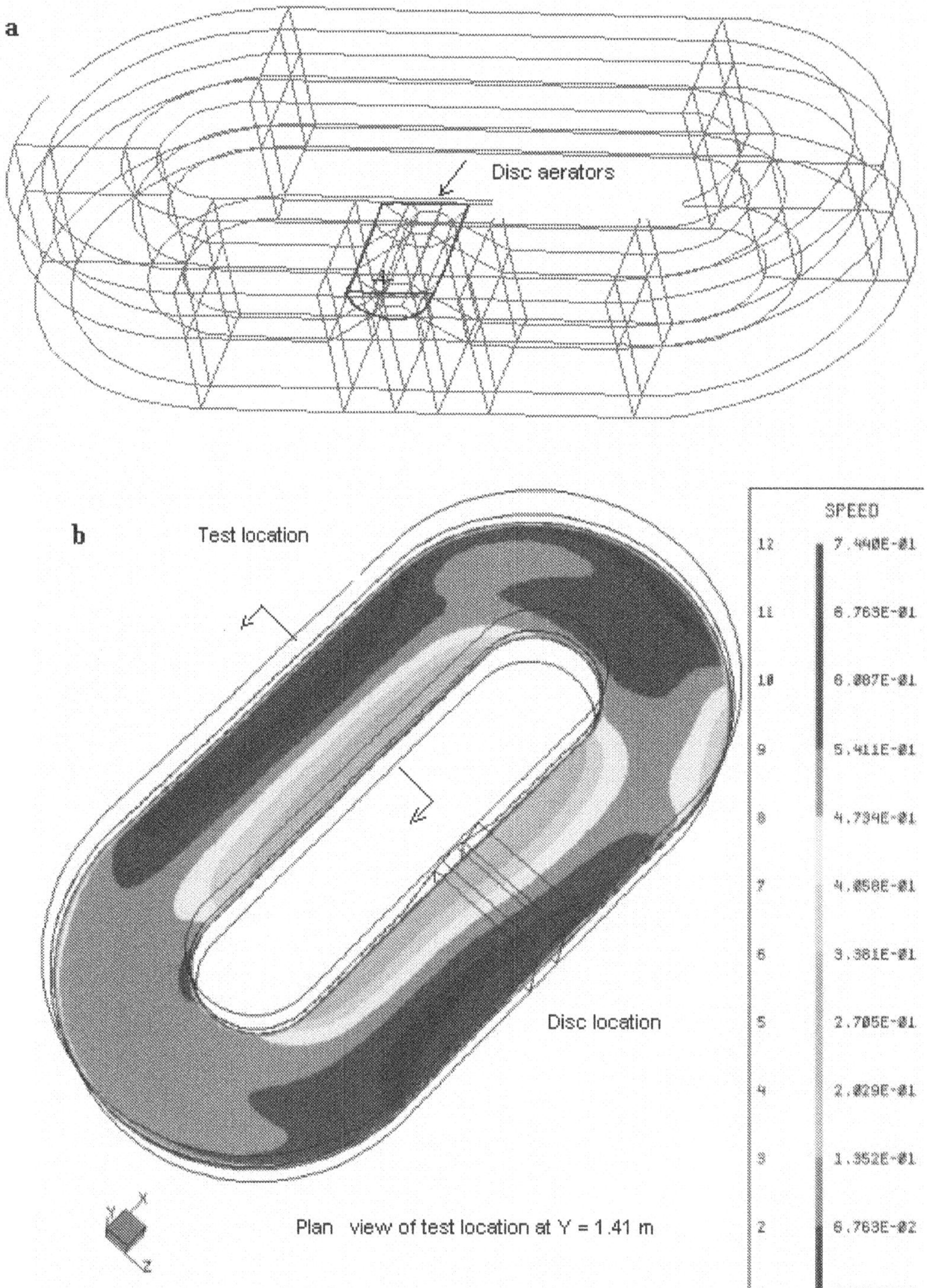

Figure 4-1. Test Tank a) Grid Structure, and b) Configuration, Test Location, and Velocity Simulation (Speed = m/s). (Reprinted with permission from Littleton et al., 2001. Color figure available on WERF website.)

4.2.1.2 Computational Fluid Dynamics (CFD) Modeling

As described in Chapter 3.0, the CFD software package used was CFX4 (ANSYS, formerly AEA Technology, Inc, Pittsburgh, PA). Fluid flow velocity was produced in the simulations by imparting a fraction of the known momentum of the fluid within the test tank, calculated based on the tank fluid flow velocity (measured) and mass, or mass flow rate across the channel cross-sectional area. For example, if the average velocity in the test tank were 0.38 m/s, and the cross sectional area were 2 m by 2 m, the mass flow rate (moving bloc) would then be 1520 kg/s (2 m x 2 m x 0.38 m/s x 10^3 kg/m^3). The calculated momentum would then be 578 kg m/s^2 (1520 kg/s x 0.38 m/s), the product of mass flow rate and velocity. A fraction of this calculated momentum source was introduced into the CFX4 geometry disc bloc structure, as illustrated in Figure 4.1. (The remainder of the observed momentum resulted from the recirculation of flow around the closed loop reactor.)

4.2.1.3 Statistical Analysis

Statistical analysis was by the method of least squares. For each value of the imparted momentum (as a fraction of the total momentum in the test tank), the velocity values determined by CFD analysis at each of nine measurement locations were compared with the test data. The residual sum of squares was then calculated as follows: 1) the residual (difference between the measured and predicted velocity value) was calculated at each measurement location, 2) this value was squared, and 3) the squared residuals for all measurement locations were summed. The residual sum of squares was calculated for a range of imparted momentum factors and the results plotted to determine the minimum. This input momentum represented the calibrated value that most accurately simulated the flow pattern in the test tank. This procedure was repeated for all the disc rotational velocities (29, 43, 55, and 60 rpm) to determine the calibrated input momentum for each. The residuals were also plotted for the selected input momentum to determine whether they were randomly distributed.

4.2.2 Full-Scale Tank

4.2.2.1 Model Development and Calibration

The CFD model for the full-scale tank was developed using the same procedures as for the test tank. In this case, the tank was the outer channel of the full-scale bioreactor at the Elmwood plant, Evesham, NJ. The performance of this bioreactor has been previously characterized (Daigger and Littleton, 2000, 1999; Littleton et al, 2000; this report, Chapter 2.0). The full-scale, closed-loop bioreactor is a three concentric channel circular tank with a total diameter of 42.1 m (138'). Each channel is 6.1 m (20') wide, with a side water depth of 3.66 m (12'). The outer channel was simulated because prior analysis demonstrated that this is where most of the SBNR occurs in this system. Normal operating conditions in the outer channel are at 45.7 to 47.8 cm (18-18.8") of disc submergence. There are two sets of disc aerators on opposite sides of the tank. Both sets generally run at 29 rpm during the summer, but only one set usually operates under winter conditions. Each set consists of 20 discs that are 137 cm (54") in diameter. Two tanks were tested, one operated with one aerator set and the other with two aerator sets. Both of them were tested at 29 rpm. An additional operational condition with higher oxygen demand in the summer was tested with two aerator sets, one at 29 rpm and the other at 43 rpm.

Cross-sectional velocity and DO concentration in the full-scale facility were measured at four locations (shown later in Figure 4-6) at water depths of 0.91, 1.8, and 2.7 m (3, 6, and 9') in the middle of the channel and 1.2 m (4') in from each sidewall (nine measurements per location).

A one-dimensional Marsh McBirney portable flow meter (model No. 2000, Milwaukee, WI) and a YSI (model 59, Yellow Springs, OH) portable DO meter were used. The membrane of the DO meter was freshly changed and calibrated before the measurements. The velocity probe was carefully oriented perpendicular to the flow during measurement, and the velocity values reported are the average of five to seven readings.

The CFD model for the full-scale tank was calibrated with the same statistical analysis approach used for the test tank. The velocity data considered in the analysis were for the cross section at location 1 (see Figure 4-6), located 3.35 m before an aerator set, which thus experienced less turbulence and variation than other locations.

4.2.2.2 Dissolved Oxygen Concentration Modeling

Oxygen source and sink terms were added to the CFD model for the full-scale tank to determine whether the heterogeneous mixing pattern could result in variations in the macroenvironment in the outer channel of the full-scale bioreactor. Previous modeling of the Elmwood plant (Daigger and Littleton, 2000, 1999; this report, Chapter 2.0) had suggested that the average oxygen requirement met in the outer channel was about 200 g/m^3-day. The total transferred oxygen mass (source term) was computed based on this volumetric rate and the total volume of the tank. A saturation factor, expressed as (10.0 -DO) / (10.0-1.0), was included. For instance, if tank DO was 1.0 mg/L, then total oxygen source = 200g/m^3-day x 2800 m^3 x saturation factor = 560 kg/day. This total oxygen source was then divided equally between the two sets of disc aerators. The oxygen-consuming reactions occurring in the bioreactor (sink term) were simulated as a uniform rate with a saturation function, as:

$$OUR = \frac{DO}{(DO+1)} \times 250 g/m^3 \text{ - day}$$

where OUR is the volumetric oxygen uptake rate and DO is the dissolved oxygen concentration at any point in the bioreactor. The maximum volumetric oxygen consumption rate with this expression approaches 250 g/m^3-day (somewhat above the approximate amount of overall oxygen input for the full-scale plant), while the half saturation constant (the DO concentration where the oxygen consumption rate is 0.5 x 250, or 125 g/m^3-day) is 1 mg/L.

The oxygen source and sink terms, as well as momentum, were implemented using the FORTRAN scalar subroutine in CFX4 called USRSRC (Anon., 2000). The system geometry was defined as the source regions (blocs containing the aeration discs) and the tank region (the rest of tank, excluding the disc blocs). Then the momentum source, oxygen source, and oxygen sink terms were added to the source region, while only the oxygen sink term was added to the tank region.

4.3 Results

4.3.1 Test Tank Velocity

Table 4-1 presents the measured velocities in the test tank at each of the nine points for each of the four disc rotational velocities. Figure 4-1b shows the simulation results for this tank at 1.41 m below the surface. Figure 4-2 presents the residual sum of squares for the measured velocity minus the predicted velocity data as a function of the input momentum (expressed as a fraction of the momentum in the test tank) for the four different disc rotational velocities. The minimum residual sum of squares occurred at a fractional momentum input of approximately 0.35 for all tested disc rotational velocities. Figure 4-3 presents the residuals for the optimum fractional momentum input for each disc rotational speed at the nine measurement points. The optimum CFD model tended to slightly under-predict (positive values in Figure 4-3) the velocity along the inner wall and in the upper portion of the test tank and to over-predict (negative values) velocity towards the outer wall and in the lower portion of the test tank. This indicates that some systematic error may have occurred in the CFD, as discussed later. The residuals ranged from <1-26% of the average test tank velocity for all disc rotational speeds.

Table 4-1. Test Tank Velocities (m/s) at Various Aerator rpm.

Distance from inner wall (in)	Measured			Calculated*		
	Top	Middle	Bottom	Top	Middle	Bottom
		at 29 rpm			Predicted with 0.35 Momentum	
15	0.26	0.18	0.16	0.204	0.166	0.145
33	0.30	0.28	0.25	0.261	0.249	0.304
60	0.32	0.35	0.40	0.347	0.367	0.421
Average		0.278 (m/s)			0.274 (m/s)	
		at 43 rpm			Predicted with 0.34 Momentum	
15	0.36	0.24	0.26	0.279	0.229	0.200
33	0.40	0.37	0.33	0.357	0.342	0.412
60	0.46	0.50	0.52	0.474	0.501	0.576
Average		0.382 (m/s)			0.380 (m/s)	
		at 55 rpm			Predicted with 0.35 Momentum	
15	0.40	0.35	0.20	0.313	0.258	0.225
33	0.46	0.45	0.38	0.401	0.384	0.461
60	0.50	0.52	0.57	0.533	0.563	0.647
Average		0.426 (m/s)			0.421 (m/s)	
		at 60 rpm			Predicted with 0.34 Momentum	
15	0.41	0.32	0.24	0.327	0.269	0.235
33	0.49	0.43	0.39	0.419	0.401	0.481
60	0.53	0.57	0.65	0.557	0.588	0.675
Average		0.448 (m/s)			0.439 (m/s)	

* Momentum = 360 kg-m/s^2

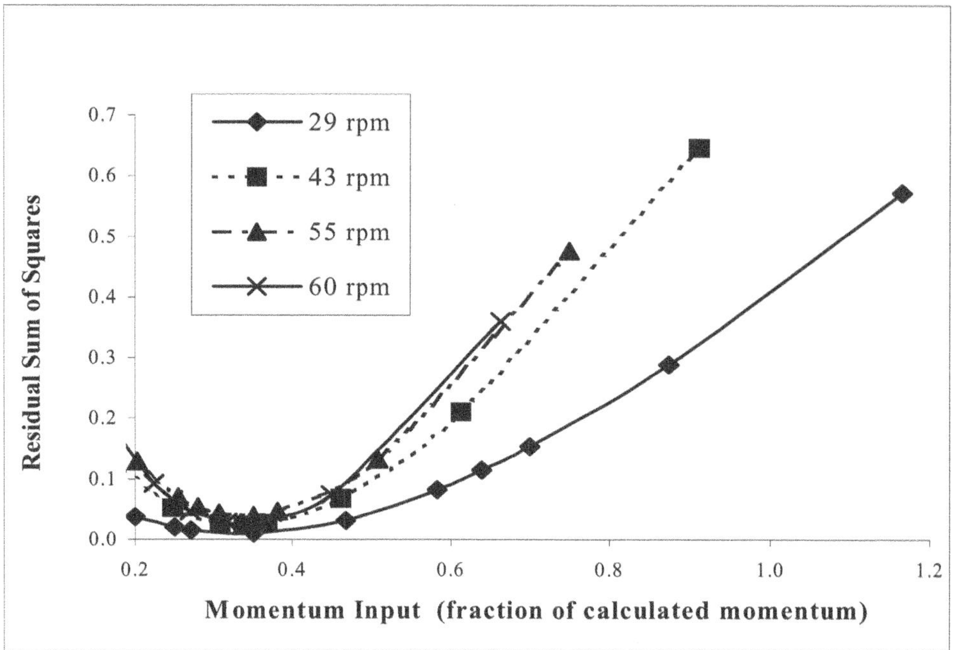

Figure 4-2. Effect of Momentum Input on Test Tank Velocity Residuals. (Reprinted with permission from Littleton et al., 2001.)

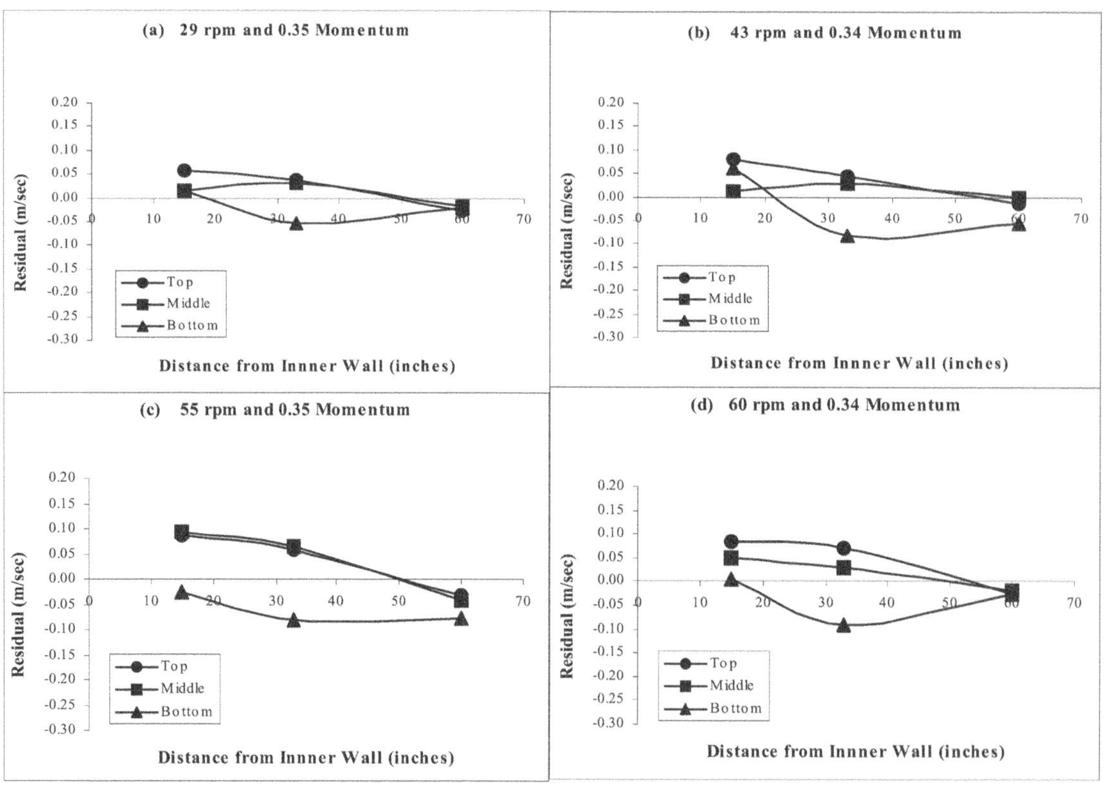

Figure 4-3. Test Tank Velocity Residuals with Optimum Momentum Input at Various rpm. (Reprinted with permission from Littleton et al., 2001.)

Figure 4-4 shows the velocity profiles predicted at various cross-sections in the test tank. Figure 4-5 shows the flow pattern and velocity distributions in the test tank predicted for a disc rotational velocity of 43 rpm at a 0.35 fractional momentum input. The "streaklines" were traced for the top layer of the cross-section of the disc aerator set, in the counter-clockwise (flow) direction. In order to more clearly show the flow pattern, only about one and a quarter revolutions are traced for each streakline. It can be seen that higher velocity flow in the top layer of the disc section went down to the bottom of the channel on the opposite side of the tank after going around the curve, then rose again after going through the second curved section. This flow pattern in the traced streaklines is consistent with the velocity profiles in Figure 4-4. All the simulations showed a similar spiral pattern.

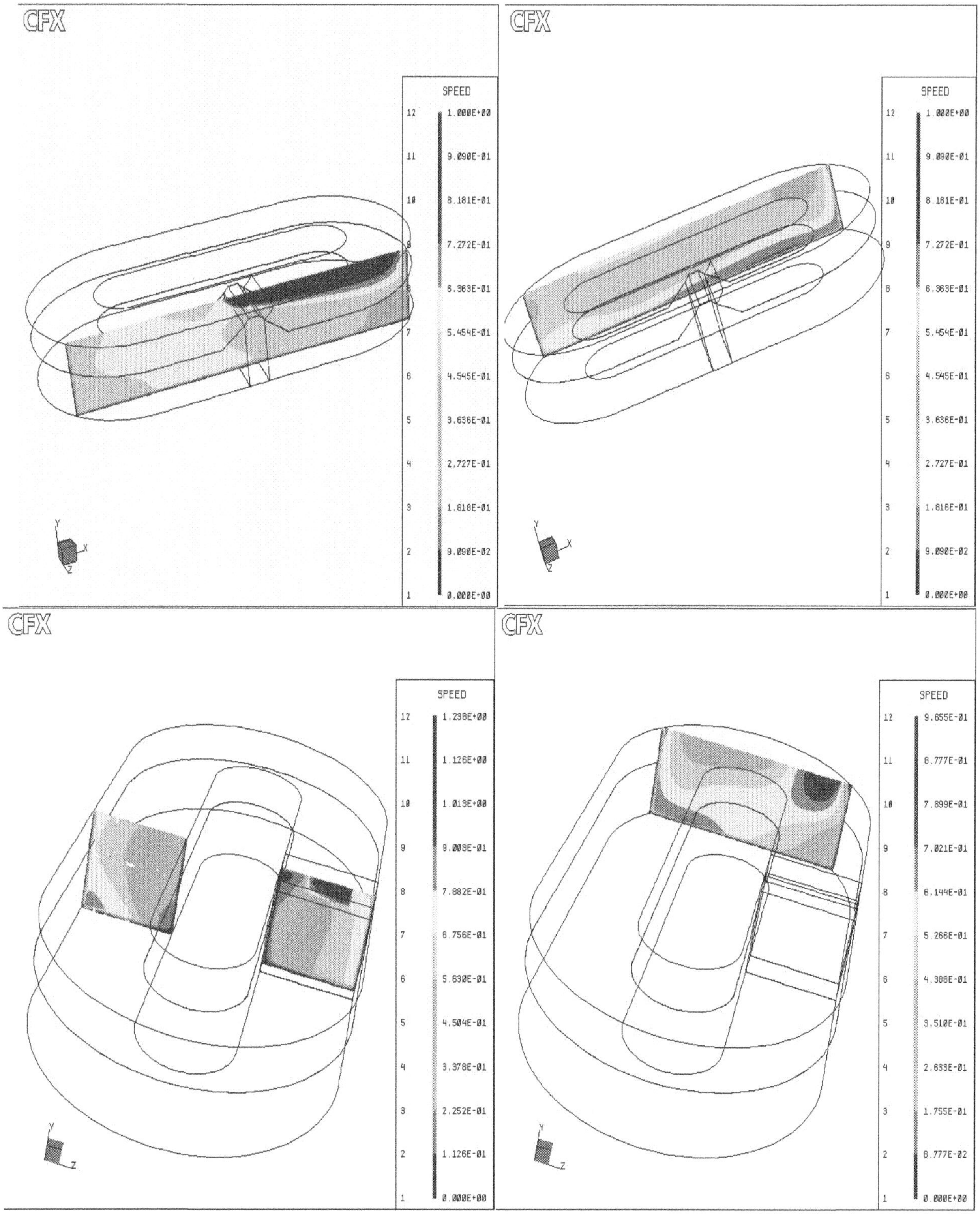

Figure 4-4. Test Tank Velocity Profile Predicted at Various Cross Sections. (Reprinted with permission from Littleton et al., 2001. Color figure available on WERF website.)

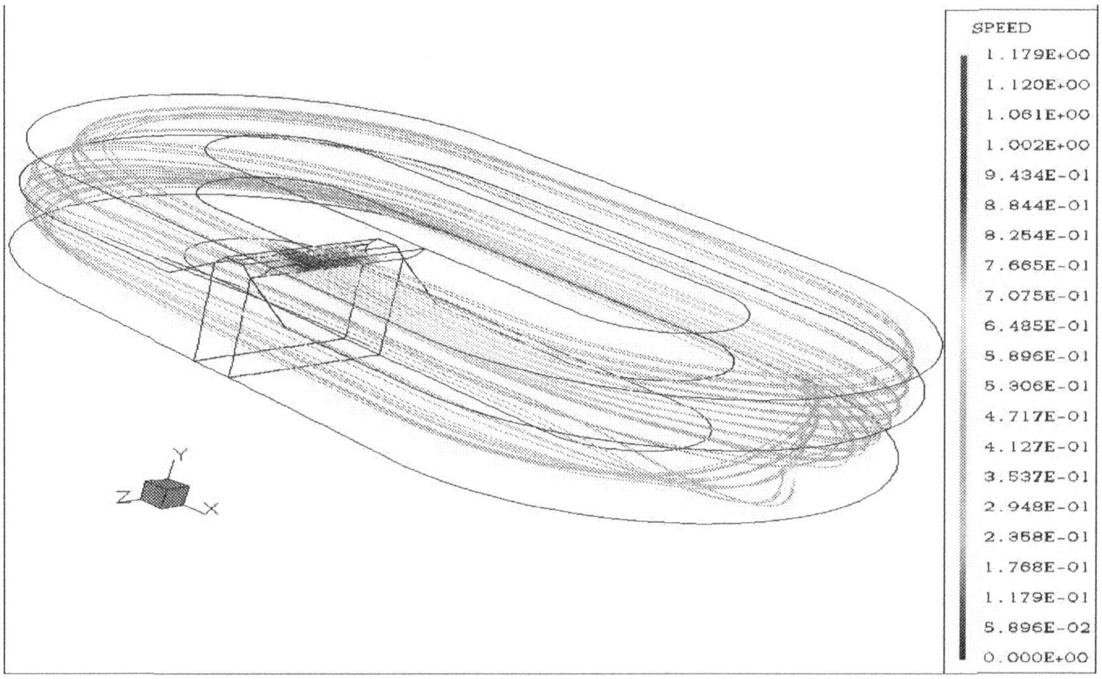

Figure 4-5. Test Tank Streaklines Showing Spiral Type Flow. (Reprinted with permission from Littleton et al., 2002b. Color figure available on WERF website.)

4.3.2 Full-Scale Tank Velocity

Table 4-2 shows velocity values measured at location 1 for three operating conditions—with one set of disk aerators operating and with two sets of disc aerators operating at two combinations of disc rotational speed. The average velocities in the full-scale tank were greater than in the test tank. In general, both measured and predicted flow velocity profile data indicate that the velocity increases towards the outer wall and decreases towards the bottom. The velocity distribution seems to be related to tank size and configuration, disc submergence, and disc rotational speed.

Table 4-2. Full-scale Tank Velocities (m/s) with One and Two Aerator Sets.

Distance from inner wall (ft)	Measured velocity at location 1			Predicted velocity (imparted momentum = 0.4 x calculated M)		
	Top	Middle	Bottom	Top	Middle	Bottom
	2 aerators (locations 1 and 3)			11,360 kg-m/s²		
	both operated at 29 rpm			each aerator imparted 5,680 kg-m/s²		
4	1.17	1.06	1.01	1.174	1.056	0.986
10	1.39	1.24	1.20	1.404	1.294	1.218
16	1.47	1.44	1.43	1.524	1.490	1.463
Average	1.27 (m/s)			1.29(m/s)		
	2 aerators (locations 1 and 3) location 1, operated at 43 rpm; location 3, operated at 29 rpm			12,700 kg-m/s² location 1 imparted 7,000kg-m/s²; location 3 imparted 5,700 kg-m/s²		
4	1.43	1.35	1.33	1.521	1.456	1.424
10	1.59	1.54	1.48	1.623	1.603	1.598
16	1.66	1.68	1.69	1.684	1.698	1.709
Average	1.53 (m/s)			1.59 (m/s)		
	1 aerator operated at 29 rpm			4,413 kg-m/s²		
4	0.83	0.78	0.75	0.751	0.692	0.642
10	0.87	0.87	0.80	0.838	0.811	0.794
16	0.86	0.87	0.84	0.932	0.931	0.931
Average	0.83 (m/s)			0.814 (m/s)		

Figure 4-6 shows the calculated velocities as well as the locations where velocity was measured. Figures 4-7 and 4-8 present traced streaklines for two and one aerator sets, respectively, in the full-scale tank as predicted by the CFD model. As with the test tank (Figure 4-5), it is clear that flow around curves generated a secondary spiral flow pattern.

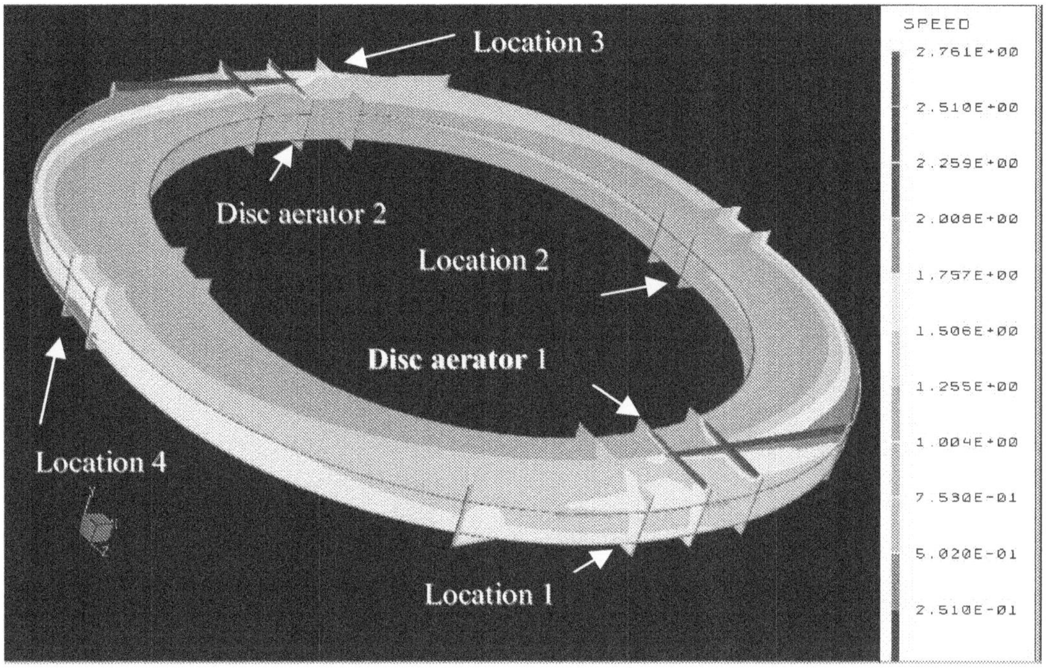

Figure 4-6. Full-scale Tank Velocity Profile with Two Aerator Sets at 29 rpm. (Reprinted with permission from Littleton et al., 2002b. Color figure available on WERF website.)

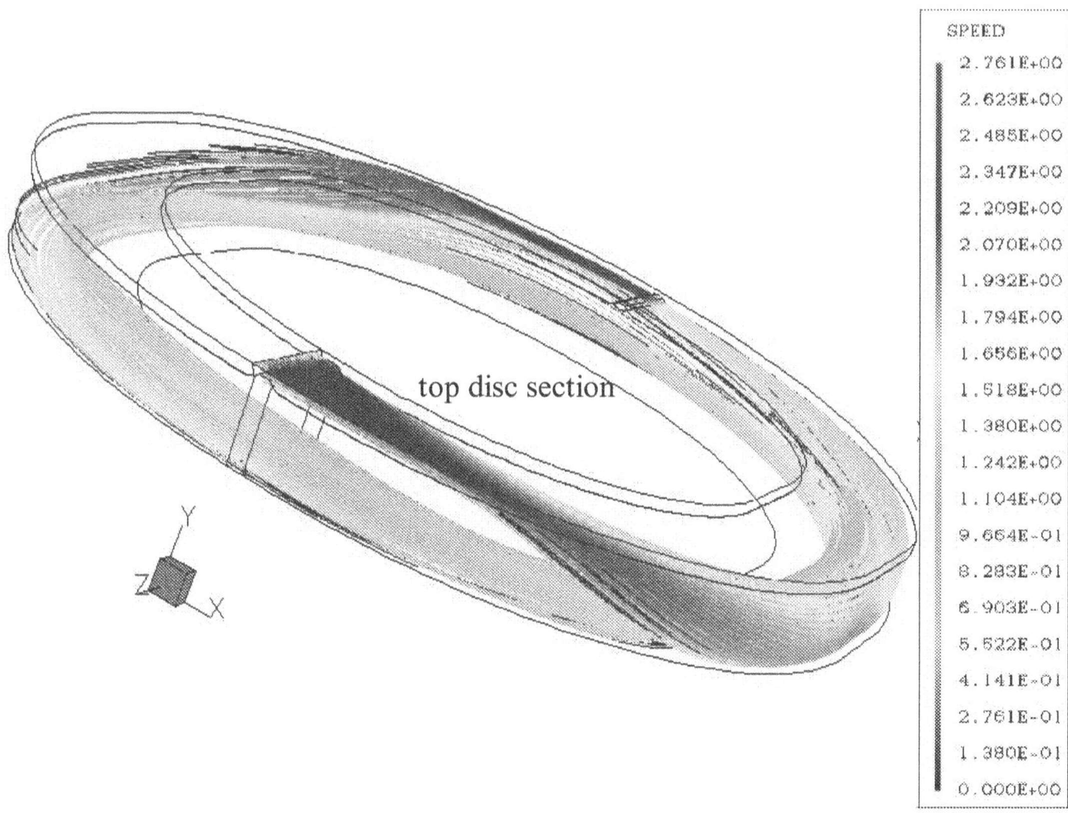

Figure 4-7. Full-scale Tank Streaklines with Two Aerator Sets at 29 rpm (Reprinted with permission from Littleton et al., 2002b. Color figure available on WERF website.)

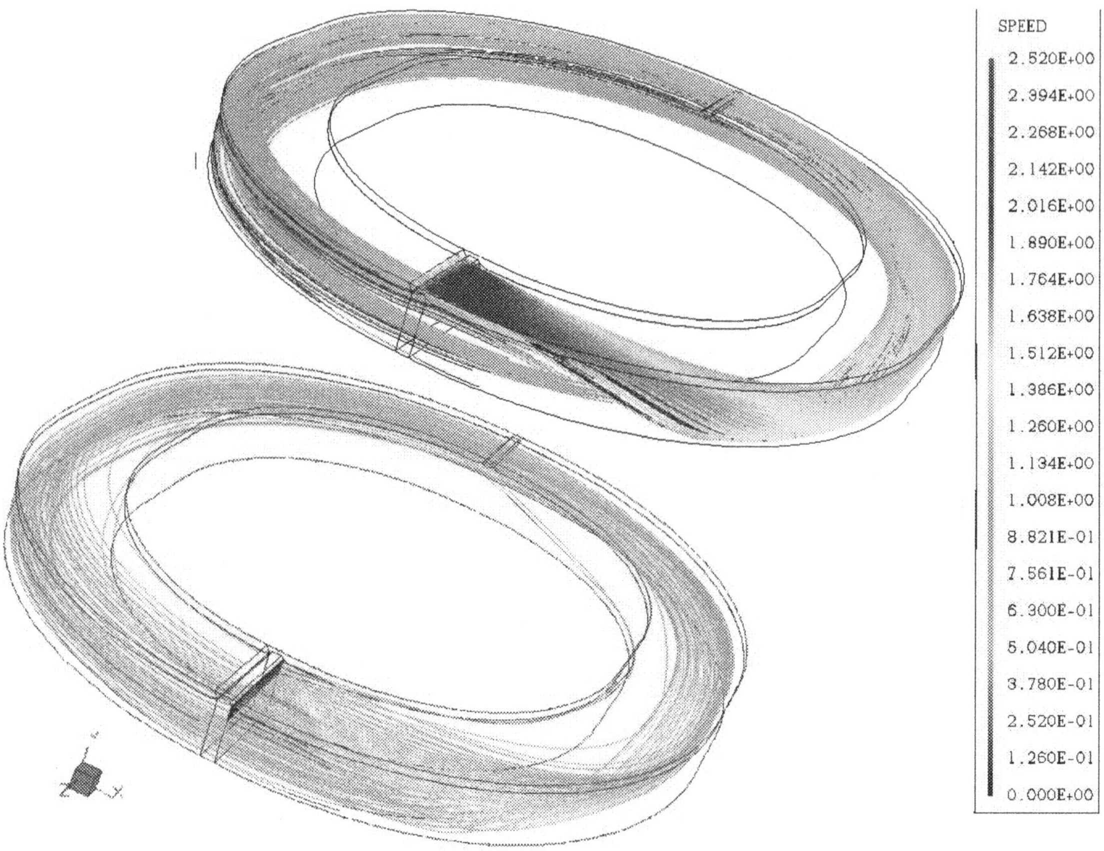

Figure 4-8. Full-scale Tank Streaklines with One Aerator Set at 29 rpm. (Color figure available on WERF website.)

Figure 4-9 presents the residual (measured minus predicted) sum of squares for the velocity data as a function of the input momentum for two operating conditions for the full-scale tank. The minimum residual sum of squares occurred at a fractional momentum input of about 0.40, which was similar to the test tank results (0.35) for the same disc rotation speed of 29 rpm.

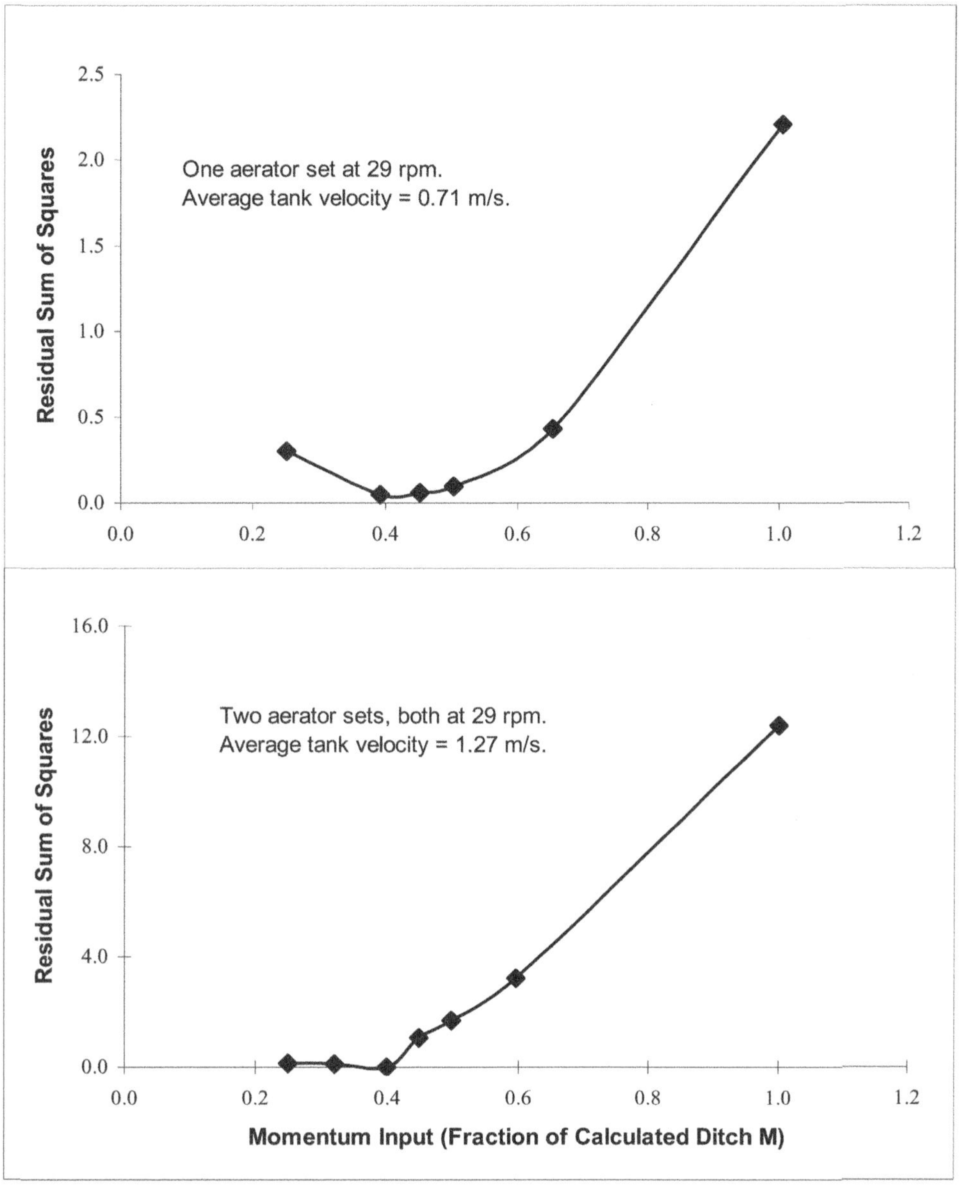

Figure 4-9. Effect of Momentum Input on Full-scale Tank Velocity Residuals.

Figure 4-10 presents the residuals for the optimum fractional momentum input for two full-scale tank operating conditions (one and two sets of disc aerators operating at 29 rpm) as a function of the velocity measurement point. Velocities in the full-scale tank were generally under-predicted along the inner wall and over-predicted at the outer wall for the one disc aerator set operating condition. The predicted velocities along the middle of the channel, from top to bottom, were generally close to the measured values. The range of residuals was about from <1 to 13%. The two aerator sets residual plot showed an even better fit (0.3-4%), and residuals seemed more randomly distributed.

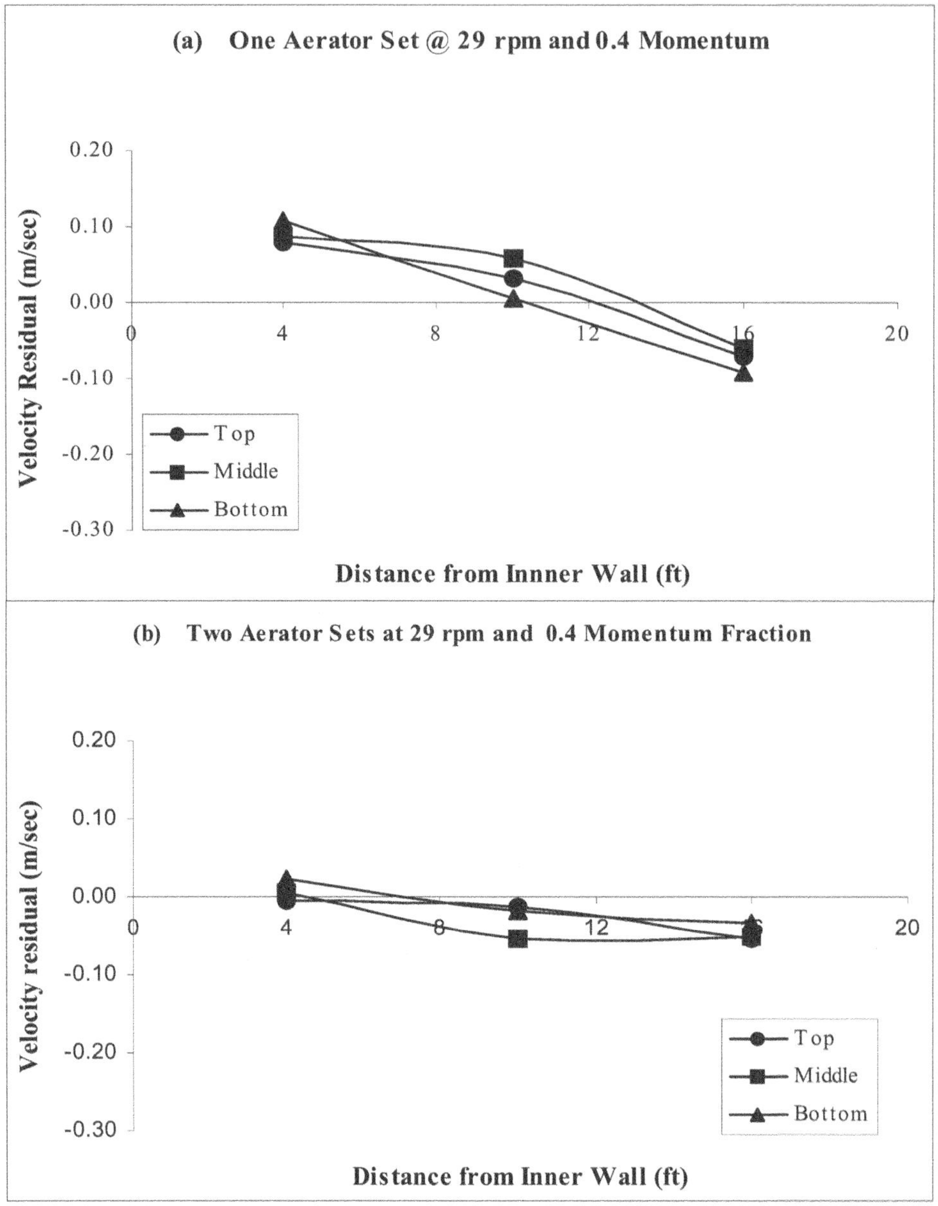

Figure 4-10. Full-scale Tank Velocity Residuals with Optimum Momentum Input for One and Two Aerator Sets.

4.3.3 Full-scale Tank Dissolved Oxygen Profile

When oxygen source and sink terms were added to the CFD simulation, a DO distribution profile was generated. Figure 4-11 presents the DO concentration profiles predicted by the simplified oxygen source and sink model with an oxygen input of 200 g/m^3-day for the full-scale tank. Higher DOs were observed near the outer wall than near the inner wall. For this simplified example, the distribution of DO was considered important, rather than the absolute values. The purpose was to test whether the mixing pattern could lead to DO gradients. The results show that the predicted DO concentration distribution is similar to the velocity profile (see Figure 4-6). Concentration gradients generated here include both convection and diffusion, along with oxygen sources and sinks.

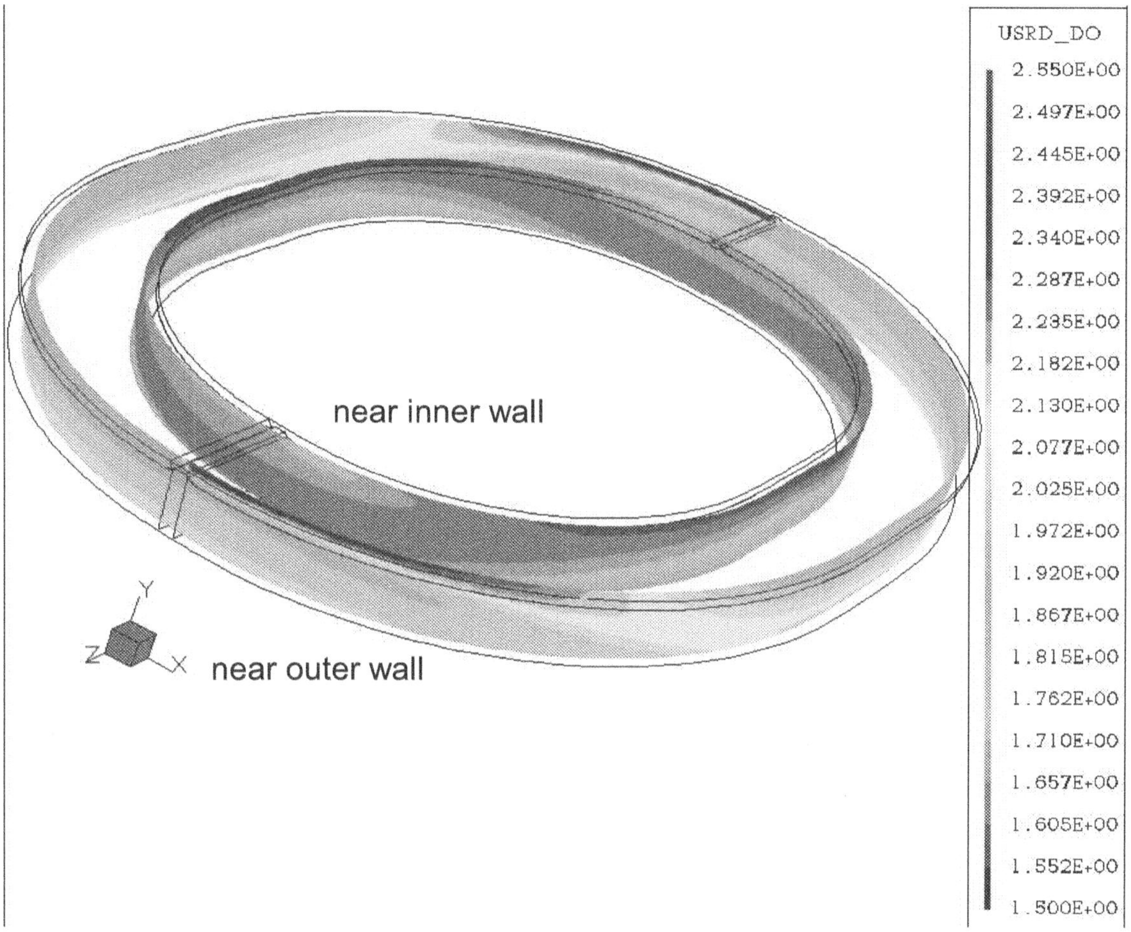

Figure 4-11. Predicted DO with Simplified Oxygen Source and Sink Terms Added. (Reprinted with permission from Littleton et al., 2002b. Color figure available on WERF website.)

The CFD results show that DO concentration gradients can develop in the tank. This is consistent with the measured DO data in Table 4-3 and observed DO in other Orbal™ plants. While there are some noticeable differences between the measured and predicted DO values in Table 4-3, they do reflect the same general trends of DO distribution. In general, higher DOs appeared towards the top of the outside of the channel and lower DOs toward the bottom of the inside of the channel.

Table 4-3. Full-scale Tank DO (mg/L) with One and Two Aerator Sets.

Distance from inner wall (ft)	Measured DO at location 1			Predicted DO simplified DO model*		
	Top	Middle	Bottom	Top	Middle	Bottom
	One aerator at 29 rpm; water surface = 3.71 m (12 ft-2 in)					
4	0.03	0.03	0.03	0.058	0.056	0.054
10	0.03	0.03	0.04	0.088	0.088	0.097
16	0.04	0.04	0.04	0.124	0.120	0.125
Average	0.034 (mg/L)			0.090 (mg/L)		
	Two aerator sets, both at 29 rpm; water surface = 3.66 m (12 ft)					
4	0.04	0.04	0.04	0.443	0.438	0.454
10	0.05	0.05	0.06	0.513	0.512	0.528
16	0.05	0.06	0.08	0.576	0.587	0.612
Average	0.052 (mg/L)			0.518 (mg/L)		

* Oxygen input of 180 g/m3-day

4.4 Discussion

The test tank results demonstrate that the developed CFD model qualitatively captured many aspects of the flow pattern in the test tank. The velocity residual sums of squares were reduced to minimum values when the imparted momentum source at the disc aerators was equal to about 35% of the calculated momentum in the test tank for all the tested disc aerator rotational velocities (29, 43, 55, and 60 rpm) (Figure 4-2). The remaining ~65% of the observed momentum presumably was from the recirculation of the liquid in the closed loop reactor.

Although agreement between model predictions and measured velocities was close, the pattern of the residuals indicated (Figure 4-3) that the model systematically tended to slightly under-predict tank velocities near the inner wall and in the upper portion of the tank, and to over-predict tank velocities near the outer wall and in the lower portion of the tank. The action of the disc aerators was simulated by simply adding a horizontal momentum source to the CFD model at the location of the disc aerators. This simplified approach was taken rather than to physically simulate the rotation of the disc aerators and the interaction of the disc aerator with the fluid in the test tank. In this work, CFD was used as a tool to provide insight into the flow pattern in the tank. In general, lack of fit within 10% is considered a good simulation for CFD applications, and within 20% is considered acceptable (Anon., 2002). The model can be refined with finer grids, and the way momentum is imparted can be modified (e.g., consideration of centrifugal force and vertical momentum at higher rotational speeds). In part, accuracy is a function of the complexity of the model, which involves a trade-off with simulation time.

Earlier simulations attempted to directly model the motion of the disc aerators and their interaction with the fluid in the test tank. However, the approach could not be carried out realistically at that time because fundamental knowledge was lacking about the interaction between the disc aerator and the fluid. The disc aerators have protrusions that create resistance to fluid motion and help pump fluid past the rotating disc. Because it was difficult to accurately simulate the effects of these protrusions, the researchers chose the simplified approach described. It seems likely that this simplification (a uniform horizontal momentum source across the disc section) is the cause of much of the lack of fit observed in Figure 4-3. One way to further improve the model's predictive capability is to refine the method for causing fluid motion in the tank.. One tactic would be to add horizontal momentum in a tapered fashion across the width of the channel. Other more fundamental approaches also can be considered. However, the model as presently developed appears to adequately predict velocity profiles for the intended purpose here, which was to examine the reactor macroenvironment's role in SBNR.

The CFD model predicts that a heterogeneous flow pattern develops within the test tank, rather than the laminar flow pattern often assumed, as a first approximation, for a closed loop bioreactor. In particular, a higher velocity is exhibited at the surface after the aerator, which then spirals down to the bottom of the tank after going through the first turn and returns to the surface at the opposite side of the channel. The CFD model qualitatively captured this flow pattern, which can be physically observed in the test tank in the flow of air entrained by the disc aerator.

Similar results were obtained when the CFD model was applied to the full-scale tank. The velocity minimum residual sum of squares was observed at a fractional momentum of about 40% of the calculated momentum in the entire tank, which is similar to the test tank results (35%). Comparisons of measured and predicted data for the full-scale tank (Table 4-2 and Figure 4-10) also indicated that the model produced a better fit for the case with two sets of aerators than for one set. Again, the CFD model tended to slightly under-predict velocities at the inner wall of the bioreactor and to over-predict velocities at the outer wall, but the differences in residuals between the top and bottom were smaller. Note that the shape of the test and full-scale tanks are different. The test tank is oblong, similar to the shape often used with oxidation ditches, while the full-scale tank is circular. It is possible that this difference may contribute to the different residuals pattern for these two tanks, although other differences (e.g., tank size, average fluid velocity, materials of construction) might also be factors. In any case, the CFD models for both tanks demonstrated that flow is not laminar. Rather, there is a more complex flow pattern, including a distribution of flows from top to bottom and from the inner to the outer wall of the tank. In addition, the researchers observed a spiral flow pattern beginning at the discharge of the disc aerator, flowing downward to the turn in the channel, and then upward on the opposite side of the channel.

Figure 4-11 demonstrates that differences in DO concentrations can occur in these tanks as a result of the heterogeneous flow pattern, the source of oxygen provided almost entirely by the disc aerators, and the more uniform oxygen demand occurring throughout the bioreactor. These results are consistent with the hypothesis that the macroenvironment created in bioreactors such as these can contribute to SBNR. The DO concentrations included in this simulation were not intended to simulate those in full-scale bioreactors. Rather, the model indicates that spatially varying DO environments can exist within such systems, and could potentially create the periodic anaerobic and anoxic conditions necessary to allow SBNR to occur.

4.5 Summary and Conclusions

CFD models were developed for a test tank and for the outer channel of a closed loop oxidation ditch of the Orbal™ configuration. A simplified approach was used to create fluid motion in the two tanks. This consisted of imparting a fraction of the tank's observed momentum uniformly across the location of the disc aerators in a horizontal direction. Potential DO variations were evaluated by including a source of oxygen at the location of the disc aerators and an oxygen sink distributed throughout the contents of the full-scale bioreactor. The researchers drew the following conclusions from this research.

1. Observations of the flow pattern in the test tank and measurements of velocities in both the test and full-scale tank demonstrated that the flow pattern within both tanks was not uniform or laminar. Instead, there was a more complex pattern including a secondary spiral flow.
2. CFD modeling was able to characterize the flow pattern in both the test tank and the full-scale tank with a good degree of fit. Velocities were not uniform across the tank cross-sectional area. The researchers observed a complex flow pattern of a high velocity discharge from the aerator, flowing downward to the bottom of the tank at the first curve and then upward to the top of the tank on the side opposite of the aerator.
    a. The velocity predictions from the CFD model of the test tank showed a good fit with measured data, although they were systematically slightly low near the top and the inner wall, and slightly high near the bottom and the outer wall. The best fit for imparting disc aerator momentum was at a factor of ~0.35 of the total calculated momentum (with the remainder resulting from recirculation) for all the tested rotational speeds (29, 43, 55, and 60 rpm).
3. The CFD velocity predictions in the full-scale tank showed good agreement with measured data using a 0.4 factor for the imparted momentum. This is similar to the value for the momentum-imparted factor in the test tank. Comparing one set of aerators versus two sets further helped confirm that the appropriate disc momentum imparting factor was used. The residual distribution showed very good fit for the middle of the channel along the top and bottom, again with a slight under-prediction at the inner wall and over-prediction at the outer wall.
4. The researchers observed a distribution of DO concentrations when a source of oxygen at the aerators and a sink for oxygen throughout the tank were added to the CFD model. The resulting DO distribution predicted by the CFD model illustrated that zones of relatively higher and lower DO concentrations can occur in the bioreactor. These results suggest that the heterogeneous flow pattern created in the bioreactor macroenvironment can result in the anaerobic and anoxic zones that would allow SBNR to occur in such aerated systems.

CHAPTER 5.0

SIMULATION OF BIOLOGICAL PHOSPHORUS REMOVAL USING COMPUTATIONAL FLUID DYNAMICS

5.1 Introduction

This chapter investigates the effects of reactor mixing pattern on biological phosphorus removal in order to provide insights into simultaneous biological nutrient removal (SBNR). The fluid field, which was previously established (Chapter 4.0) using computations fluid dynamics (CFD) in a full-scale closed-loop bioreactor (Orbal™ oxidation ditch), was further developed for testing the hypothesis that the system contains undefined anoxic/anaerobic and aerobic zones , a condition that favors the presence of phosphorus accumulating organisms (PAOs).

This study combines biochemical and fluid dynamics models in activated sludge systems for the first time. It was designed to provide insights on how chemical compounds react in the fluid field. The main focus of this phase of the research was to provide insights into SBNR better understanding how the macroenvironment created by the mixing patterns in closed-loop bioreactors affects PAO and general heterotroph competition,. In the research presented here, biochemical equations representing growth and decay of heterotrophs and PAOs were added to the previously developed CFD simulations, which had been verified with velocity measurements and simulations using the International Water Association's Activated Sludge Model No. 2 (ASM2; Henze et al., 1995). Results were compared to simulations using ASM2 and an analytical solution. The oxygen input was varied to determine whether this could lead to 1) dominance of heterotrophs under high oxygen conditions, 2) PAOs under lower oxygen input conditions, and heterotrophs and 3) PAOs coexisting in a heterogeneous environment at intermediate inputs, as can occur in the full-scale plants studied.

5.2 Materials and Methods

5.2.1 Modeling Approach

The first step of the approach was to adapt the previously established CFD fluid field, including the optimum imparting of momentum based on the velocity calibration and disc information for the outer channel of the Elmwood Orbal™ wastewater treatment plant Evesham, NJ (Littleton, et al., 2001; Chapter 4). Then the researchers simulated the growth of heterotrophs and PAOs, alone and in competition, with varied oxygen inputs. Both the International Water Association's Activated Sludge Model No. 2 (ASM2, Henze et al., 1995) and CFD were used separately in order to examine the reliability of the CFD model. Finally, CFD was used to test the hypothesis that the heterogeneous environment predicted in the outer channel could allow PAOs to grow and remove phosphorus, as had been observed in the full-scale plant.

5.2.2 CFD Software and Simulated Reactor Set Up

CFD utilizes numerical algorithms that can tackle fluid flow problems. Commercial CFD packages include sophisticated user interfaces to input the system's parameters and to examine the results. The software used in this research was CFX4.4, a product of ANSYS (formerly AEA Technology Inc, Pittsburgh, PA).

The CFD mixing model (Littleton et al., 2001; Chapter 4) was developed for the outer channel of one of the full-scale Elmwood Orbal™ oxygen ditches, Evesham, NJ. It**(AB asks: does "it" refer to the ditch or the channel?)** is 42.1 m (138') in diameter, 6.1 m (20') in channel width, and 3.66 m (12') in water depth. The researchers used a solids retention time (SRT) of four days (175,000 gal/day inflow to a 700,000 gal reactor, which is the outer channel volume). Since there was no return sludge in the model system, acetate and phosphate were fed in the influent at concentrations of 1500 and 150 or 50 mg/L, respectively.

5.2.3 ASM2 Modification and Simulations

The International Water Association's activated sludge models have been carefully developed and well tested, and hence widely adopted for the activated sludge process. ASM2, along with the analytical solution for aerobic heterotrophs alone, was used to validate the mass balances for compounds in the CFD simulations. The heterotroph and PAO simulations were conducted separately for an aerobic, single chemostat reactor with DO set at values ranging from 0.01 to 8.0 mg/L (Figure 5-1a).

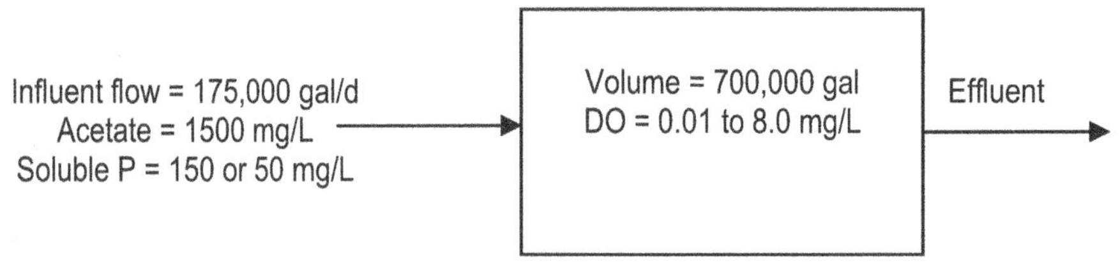

Figure 5-1a. Schematic for Single Completely Mixed Reactor.

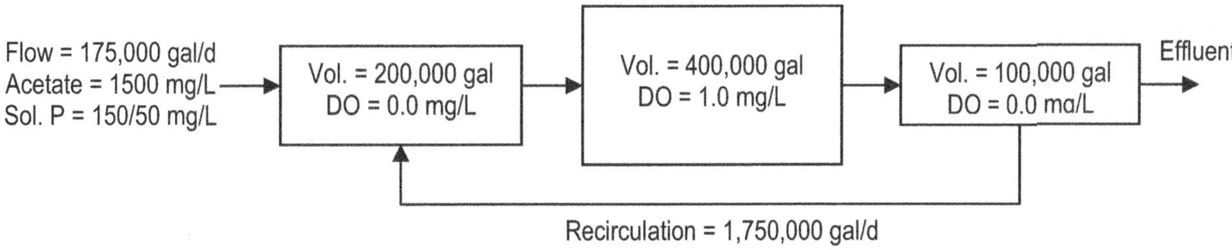

Figure 5-1b. Schematic for Completely Mixed Anaerobic/Aerobic/Anoxic Reactors in Series.

The competition of heterotrophs and PAOs growing together also was simulated, both in a single reactor (Figure 5-1a) and in an anaerobic/aerobic/anoxic reactor series with internal recirculation (Figure 5-1b). Both of these configurations were set up as completely mixed reactors (in series for Figure 5-1b) with a total SRT of four days. The DO concentration in the single reactor and in the aerobic compartment of the reactor series was varied from 0.01 to 8.0 mg/L. The researchers also examined competition of heterotrophs and PAOs at various aerobic SRTs by varying the relative aerobic and anaerobic volumes.

ASM2 contains a total of 20 kinetic expressions for a variety of species. The reactions here only focused on growth and decay of the general aerobic heterotrophic populations and the PAOs. Thus ASM2 was modified to reflect this simpler system. The modified model includes the three soluble species which include dissolved oxygen (DO, S_O), acetate (S_{AC}), and phosphate-phosphorus (S_P); four particulate species representing the masses of PAOs (X_{PAO}), polyphosphate (X_{PP}), poly-β-hydroxyalkanoates (X_{PHA}), and heterotrophic biomass (X_{het}); and eight process rate expressions. In addition, the traditional decay model was used to replace the lysis model of ASM2 in order to capture the oxygen consumption behavior of these two microorganisms. The stoichiometry of these components and the rate expressions used in ASM2 are presented in Table 5-1. Since the phosphorus used for growth and decay of heterotrophs and PAOs has little effect on the competition between these two microorganisms, these terms (shown with * in Table 5-1) were not included in the models, in order to simplify the simulations. Note that acetate and biomass concentrations are expressed on a chemical oxygen demand (COD) basis.

Table 5-1. Model Components: Stoichiometry, Rates, and Equations with PAOs and Heterotrophs.

Process	S_O	S_{AC}	S_P	X_{PAO}	X_{PP}	X_{PHA}	X_{HET}	Rate (day^{-1})	Rate Expression
PAO PHA Storage	0	-1	0.4	0	-0.4	1.0	0	3.0	$k_r X_{PAO}\left(\frac{S_{AC}}{k_{S2}+S_{AC}}\right)\left(\frac{X_{PP}/X_{PAO}}{k_{PP}+X_{PP}/X_{PAO}}\right)$
PAO PP Storage	-0.2	0	-1.0	0	1.0	-0.2	0	1.5	$k_r X_{PAO}\left(\frac{S_O}{k_{S1}+S_O}\right)\left(\frac{S_P}{k_{S3}+S_P}\right)\bullet\left(\frac{k_{P\max}-X_{PP}/X_{PAO}}{1.05\,k_{P\max}-X_{PP}/X_{PAO}}\right)\left(\frac{X_{PHA}/X_{PAO}}{k_{PHA}+X_{PHA}/X_{PAO}}\right)$
PAO Growth	-0.6	0	-0.02*	1.0	0	-1.6	0	1.0	$k_r X_{PAO}\left(\frac{S_O}{k_{S1}+S_O}\right)\left(\frac{S_P}{k_{S3}+S_P}\right)^{*}\bullet\left(\frac{X_{PHA}/X_{PAO}}{k_{PHA}+X_{PHA}/X_{PAO}}\right)$
PAO Decay	-1.0	0	0.02*	-1.0	0	0	0	0.05	$k_d X_{PAO}$
PP Decay	0	0	1.0	0	-1.0	0	0	0.05	$k_d X_{PP}$
PHA Decay	0	1.0	0	0	0	-1.0	0	0.05	$k_d X_{PHA}$
HET Growth	-0.6	-1.6	-0.02*	0	0	0	1.0	6.0	$k_r X_{HET}\left(\frac{S_O}{k_{S1}+S_O}\right)\left(\frac{S_P}{k_{S3}+S_P}\right)^{*}\left(\frac{S_{AC}}{k_{S2}+S_{AC}}\right)$
HET Decay	-1.0	0	0.02*	0	0	0	-1.0	0.1	$k_d X_{HET}$

* P term omitted from simulations

S_O — dissolved oxygen, mg/L
S_{AC} — dissolved acetate, mg/L
S_p — dissolved phosphate, mg/L
X_{PAO} — PAO biomass in COD unit, mg/L
X_{PP} — poly-P concentration in biomass, expressed as a liquid phase P concentration, mg/L
X_{PHA} — stored PHA concentration in COD unit, mg/L
X_{HET} — heterotroph biomass in COD unit, mg/L
K_{S1} — half saturation coefficient for dissolved oxygen, = 0.2 mg O_2/L
K_{S2} — half saturation coefficient for acetate, = 4.0 mg COD/L
K_{S3} — half saturation coefficient for dissolved phosphate, = 0.2 mg P/L
K_{PP} — half saturation coefficient for poly-P, = 0.01 mg P/L
K_{Pmax} — constant for maximum value of poly-P storage, = 0.34
K_{PHA} — half saturation coefficient for PHA, = 0.01 mg PHA/mg PAO
k_r — maximum specific reaction rate for specified species, hr^{-1}
k_d — decay coefficient for specified species, hr^{-1}

5.2.4 Analytical Solution

An analytical solution for heterotrophic biomass (X_{het}) and acetate (S_{AC}) concentrations also can be calculated at a given DO (S_O), since the SRT, yield (Y_h), maximum growth rate (μ_{max}), decay rate (k_d), and half-saturation coefficients for oxygen (K_{S1}) and acetate (K_{S2}) are known (specified in Table 5-1) for the simulation. For acetate:

$$\frac{1}{SRT} = \mu_{max}\left(\frac{S_O}{K_{S1}+S_O}\right)\left(\frac{S_{AC}}{K_{S2}+S_{AC}}\right) - k_d$$

Substituting in the values from Table 5-1 and rearranging:

$$\left(\frac{S_{AC}}{4.0+S_{AC}}\right) = \frac{\frac{1}{SRT}+k_d}{\mu_{max}}\left(\frac{K_{S1}+S_O}{S_O}\right) = \frac{\frac{1}{4}+0.1}{6.0}\left(\frac{0.2+S_O}{S_O}\right) = 0.0583\left(\frac{0.2+S_O}{S_O}\right)$$

For heterotrophic biomass, with influent acetate concentration represented as $S_{AC_{in}}$:

$$X_{het} = \frac{Y_h(S_{AC_{in}} - S_{AC})}{1+k_d SRT} = \frac{\frac{1.0}{1.6}(1500 - S_{AC})}{1+0.1\times 4} = 0.446(1500 - S_{AC})$$

5.2.5 CFD Model

The CFD model is based on the fundamental governing equations of fluid dynamics — the continuity, momentum, and energy equations. CFD models typically include a description of the flow geometry, a set of coupled differential equations describing the physics and chemistry of the flow, boundary and initial conditions, and a structured mesh of points at which these equations are solved (Warsi, 1993). CFX4 uses a finite volume technique (Lesieur et al., 1996) to solve a basic set of Navier-Stokes equations for conservation of mass and momentum (Anon., 2000). There are several forms of the Navier-Stokes equations, but the most useful for the development of the finite control volume were shown in Littleton et al. (2001) and Chapter 4.

Various extensions are used for incorporating momentum, heat, or mass transport, which can be expressed as a generic scalar (nonvector) advection-diffusion equation known as the scalar transport equation:

$$\frac{\partial(\rho\phi)}{\partial t} + \nabla\bullet(\rho\mathbf{u}\phi) - \nabla\bullet(\Gamma\nabla\phi) = S \qquad 5.2.5\text{-}1$$

where S is a source or sink term representing creation or destruction of a component ϕ, (*e.g.*, momentum or DO), Γ is the diffusion coefficient, ρ = density, and $\mathbf{u}$ = (u,v,w) is the fluid velocity vector.

5.2.5.1 User Scalar Equation and Source/Sink Terms

Transport equations for additional scalar quantities, such as biomass and substrate concentrations, may be solved along with the flow equations (the default scalar value is zero in generic equation 5.2.5-1). All the species in the transport equation are handled as a mass fraction and have dimensions of mass/time. When user scalar equations are being solved, the molecular diffusivity of each species must be specified. If ϕ represents an additional species mass fraction, then $\Gamma = \rho\ D_\phi$, where D_ϕ has the usual definition of the scalar's diffusivity. When S is a source term, it must be specified as non-zero either using command language for a relatively simple source or using the CFX4.4 User Subroutine USRSRC. When USRSRC is called, the equations have been linearized and are expressed in the steady state form:

$$\text{Convection - Diffusion} = \text{Sources}$$

The system geometry was defined as two source regions (disc blocs) and a tank region (the rest of tank, excluding the disc blocs). The momentum source and oxygen source were both added in the disc region as a 3-D patch based on the known plant operation. Influent flow including acetate and phosphate were added at the inlet, which was a 2-D patch on top of the 3-D tank (an open pipe, with flow dropping to the tank). A total momentum of 11,360 kg.m/s^2 for the CFD fluid field was previously established (Littleton, et al., 2001; Chapter 4), based on the optimized imparted momentum from the velocity calibration and disc information in the outer channel. Oxygen input ranged from 20 to 1000 mg/L-day, split evenly between the two disc regions. Considering oxygen mass transfer, the oxygen source for each disc aerator was expressed as:

$$\text{Oxygen source} = \frac{\text{oxygen input}}{2}(\text{mg/l - day}) \times \text{tank } volume \times \frac{(10 - DO)}{(10 - 1)}$$

The chemical reactions (Table 5-1) were applied to the entire tank (both the tank region and the disc regions). The reaction terms for scalar mass fractions (source and sink terms) were calculated as shown below, based on the stoichiometry and rate expressions listed in Table 5-1. The unit conversion from mass fraction (M) to scalar concentration (mg/L) is mg/L = M (kg/kg) • Density (kg/L) • 10^6 mg/kg. Following is an example of source/sink addition for heterotrophs based on Table 5-1:

$$\text{Oxygen source/sink} = -0.6(6.0)X_{\text{HET}}\left(\frac{S_{\text{O}}}{k_{\text{S1}} + S_{\text{O}}}\right)\left(\frac{S_{\text{AC}}}{k_{\text{S2}} + S_{\text{AC}}}\right) -1.0(0.1)X_{\text{HET}}$$

$$\text{Acetate source/sink} = -1.6(6.0)X_{\text{HET}}\left(\frac{S_{\text{O}}}{k_{\text{S1}} + S_{\text{O}}}\right)\left(\frac{S_{\text{AC}}}{k_{\text{S2}} + S_{\text{AC}}}\right)$$

$$\text{Heterotroph source/sink} = 1.0(6.0)X_{\text{HET}}\left(\frac{S_{\text{O}}}{k_{\text{S1}} + S_{\text{O}}}\right)\left(\frac{S_{\text{AC}}}{k_{\text{S2}} + S_{\text{AC}}}\right) -1.0(0.1)X_{\text{HET}}$$

Source/sink terms for PAOs and PAO/heterotroph competition use the same approach to sum the variables for each species in the stoichiometry column, then multiply by the rate expressions according to Table 5-1.

Additional FORTRAN subroutines were used in the simulation. USRCVG was called at the end of each iteration to allow users to modify and set their own convergence criteria during the run. USRINT allows overwriting the default initial conditions and writing final calculated quantities, such as volume. USRTRN is called after the initial estimate and at the end of each time step. It can be used to monitor the calculation or to produce special output for each step to examine the progress of the solution. In our simulation it was particularly useful for examining the mass balance at the effluent boundary.

5.2.5.2 Initial and Boundary Conditions

Initial conditions can be crucial in steady-state simulations. Starting with near steady-state conditions can greatly accelerate a run by decreasing the number of iterations needed to converge on the solution. An analytical solution for heterotrophs was found using a dynamic Excel worksheet to see how initial conditions, time steps, and parameters such as K_s affected the calculations. This also gave an approximation of the steady state values. The strategy was to start with simple runs to see if a mass balance could be achieved for all species in the same reactor conditions despite the configuration differences. The ASM2 model and analytical solution gave good insight on expected results to validate the CFD results. When experimental data cannot be readily obtained, well-accepted models are often used to calibrate a CFD model. The simulation was first run with the only abovementioned known momentum source, until it achieved a steady-state fluid field. Then the hydrodynamic equations were turned off and the scalar quantities introduced for the heterotrophs-only simulation. In this way once the hydrodynamic variables converged, the biochemical simulation could be run with much less computer time. Based on the ASM2 and analytical simulation results, initial estimates for concentrations were 1.0 mg/L acetate and 670 mg/L heterotroph biomass. In order to simplify the problem, phosphate for cell synthesis was ignored at this stage. Similar procedures were applied to the PAO simulations, except that phosphate feed was introduced with an Acetate/P ratio of 10. The initial estimates for PAOs were acetate = 1.0 mg/L, P = 0.7 mg/L, PAO = 690 mg/L, PP = 150 mg/L and PHA = 5 mg/L.

The simplified bioreactor with a hydraulic residence time (HRT) of four days (also the SRT, since there was no sludge return) had an influent flow of 0.0076 m^3/s (based on an outer channel volume of 2800 m^3). The influent flow through a 12' pipe on the top of the tank had a velocity of 0.11 m/s (7.21 kg/s of mass flow), which was set in the command language. The inlet boundary was specified as a 2-D patch (as an influent flow dropping to the tank through an open pipe) over the 3-D tank. The tank walls (side and bottom) were specified as a non-slip boundary condition, which means that velocities at the boundary walls are equal to zero. The top, free surface of the liquid in the tank was specified as a slip wall, subject to the fluid velocity. The outlet was located on the inner wall near the bottom of the tank, as a 36' x 36' (0.91 x 0.91 m) square gate. It was set as a 2-D patch pressout (governed by static pressure), which is often the choice when chemical compounds are involved. The researchers applied double precision despite the longer computer run time required, because it was necessary to resolve a small flow and low scalar concentration (some species as low as 0.01 mg/L, which is 10^{-8} mass fraction unit, as used in CFX4.4) mixed with a large recirculating flow; hence, round-off error could be crucial..

5.3 Results and Discussion

If theories are only an approximation of reality, then models can be considered as the next step towards it. Reality may be infinitely complex and thus often needs to be simplified, reduced, and disintegrated in an attempt to capture it in a model. CFD can generate an unlimited choice of levels of detailed results. The purpose of computing here was insight, not numbers (Versteeg and Malalasekera, 1995; van der Walt, 2000). The main outcome of a CFD exercise is improved understanding of the behavior of a system, but since there are no cast iron guarantees of the accuracy of a simulation, the results must be validated. This was done here via comparison with ASM2, and where possible, with an analytical solution.

5.3.1 Heterotroph Simulations

Table 5-2 and Figure 5-2 compare the heterotroph simulation results for the CFD, ASM2 (single reactor), and analytical solution for acetate and biomass concentration with varied DO concentrations. Acetate COD is almost completely converted to biomass COD under aerobic conditions in a single bioreactor. The CFD results are very similar to those of the ASM2 and the analytical solution. This comparison helps to validate the mathematical model and source/sink transport equations.

Table 5-2. Separate Heterotroph and PAO Simulations.

Scalar	Heterotrophs*			PAOs **		
(mg/L)	Analytical solution	ASM2 single	CFD	ASM2 Single	ASM2 anaer./aerob./anox.	CFD
DO	1.0	1.0	1.0***	1.0	0.0/1.0/0.0	0.92***
Acetate	0.30	0.32	0.33	0.93	0.92	0.93
P	-	-	-	0.48	0.46	0.34
X_{Het}	669	665	678	-	-	-
X_{PAO}	-	-	-	697	697	689
X_{PP}	-	-	-	152	149	147
X_{PHA}	-	-	-	4.0	4.0	4.3

* Influent P = 0 mg/L ** Influent P = 50 mg/L *** DO value at middle of disc region, 8.3 inches below water surface.

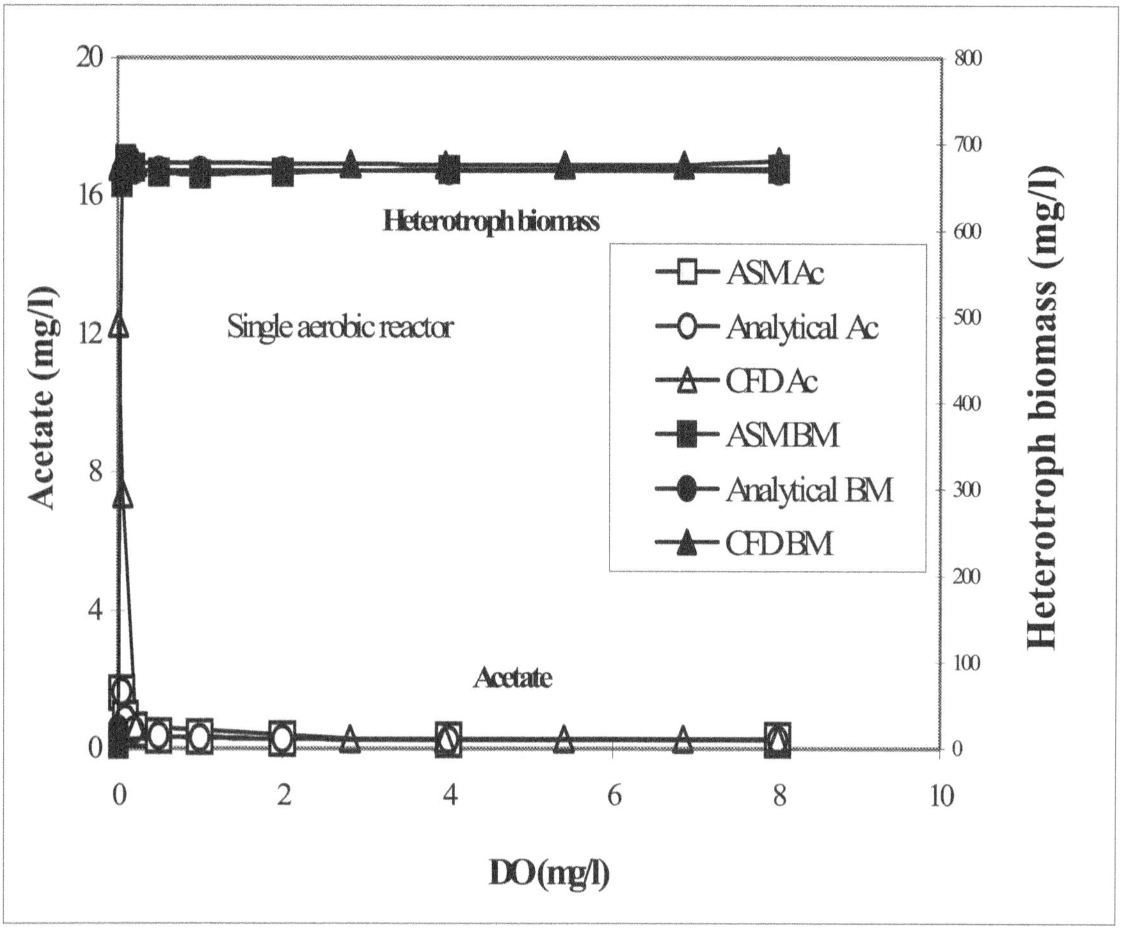

Figure 5-2. Heterotroph Simulations in CFD and ASM2 and Analytical Solutions. (Reprinted with permission from Littleton et al., 2002b.)

With simple source/sink additions, such as for heterotrophs alone, the CFD simulation converged on a solution well and reached it relatively quickly, within 3,000 iterations at higher DO inputs. At lower DO inputs the simulations slowed down somewhat but still reached steady state within 6,000 iterations. Good mass balances for all the chemical constituents were achieved within a short time (about six hours using a 1 GHz Intel Pentium IV Processor with 500 MB ROM).

One problem encountered in implementing the CFD model may be worth noting. At first, the simulations were giving reasonable results, but these showed a gradual decrease in the biomass with additional iterations leading to a growing failure to achieve a mass balance. Eventually, with the help of technical consulting from the CFX provider, the researchers discovered the problem: As the main recirculation flow moved forward past the effluent patch located on inner wall, a small amount of effluent was drawn back into the main channel. This is in fact expected, and does not change the overall fluid flow. However, as default values in the negative flow (flow out through the outlet is positive, while that back in from the outlet is negative), CFX4 assigns a zero to any dependent variables—such as DO, biomass, and substrate concentrations. In reality, such flow is expected to contain the effluent concentrations of these parameters. Thus while the output results for the velocity and mass flow rate converged, the outflow concentrations of the scalars was affected by the "drawn-in" flow assigned zero concentrations. This led to a gradual loss of mass from the system, explaining the failure to obtain a mass balance. This error was hard to detect because the influent and effluent are a very small fraction of the large recirculating flow within the tank. This problem was solved by using a FORTRAN subroutine to substitute the outlet neighboring cell concentrations for the zero default values.

5.3.2 PAO Simulations

Figure 5-3 and Table 5-2 show the individual PAO simulations in the ASM2 (single and anaerobic/aerobic/anoxic reactors) and CFD models. PAOs alone grow in all three cases. Again, in similar DO environments, mass balance was maintained, and there was good agreement between the models for all the chemical species. However, with more variables and the greater range of concentrations of species (and hence, increased complexity), the CFD model had more calculation problems with divergence and mass tolerances. Solutions were possible by adding an under relaxation factor (URF), or applying a more sophisticated solver allowed for solution, but the trade-off was greater computing time. Although the simulations were not running in dynamic conditions, a pseudo-time evolution step, such as the URF, could be applied to steady-state simulations. When the transport equation is discretized the variables can develop non-linearity, causing instability difficulties (divergence). A small URF will lead to a faster solution of linear equations and will also yield a more diagonally dominant matrix, which will minimize the non-linearity. Hence a more accurate solution can be achieved sooner.

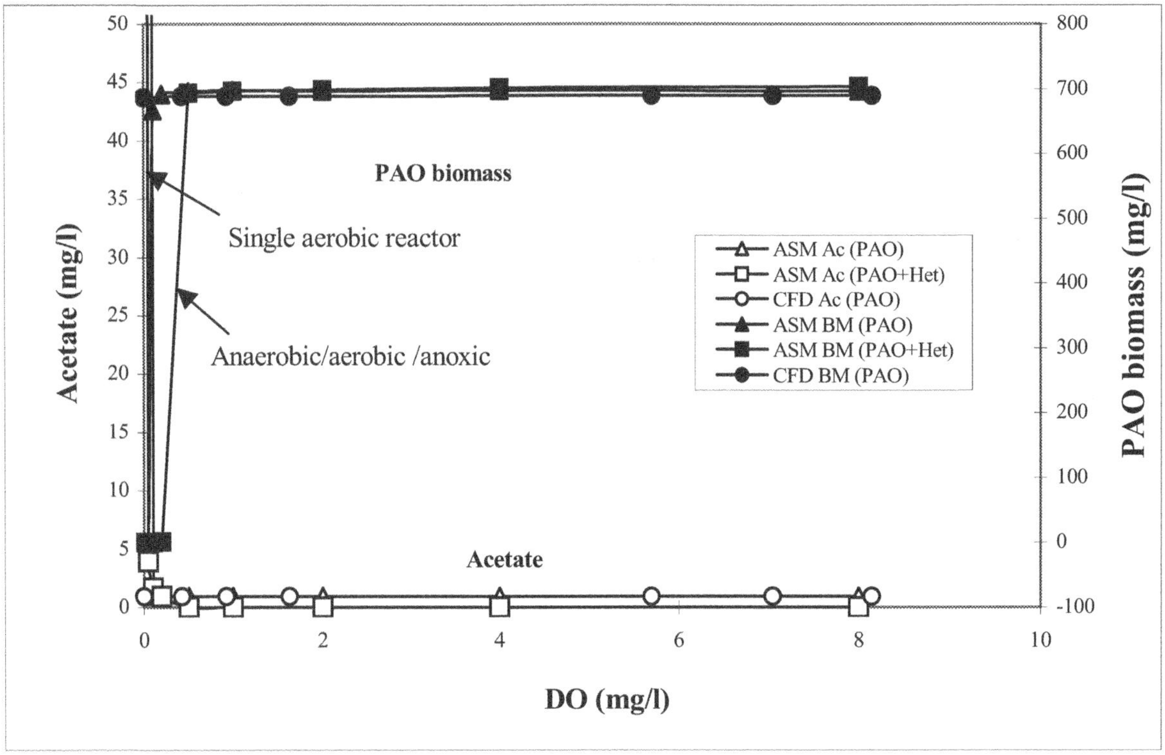

Figure 5-3. PAO Simulations in ASM2 and CFD. (Reprinted with permission from Littleton et al., 2002b.)

Again, the numerical difficulty was greater at lower oxygen inputs, where DO concentrations from about 0.05 to 1.0 mg/L resulted. Since all scalars (expressed in mg/L) were calculated from the mass fraction (with 0.05 mg/L representing a mass fraction of 5×10^{-8} kg/kg), double precision was necessary to prevent round-off error, again slowing down the simulation run time. However, with the still relatively simple source/sink terms for PAOs alone, simulations reached steady state within 3000 iterations, although each iteration took longer than for the heterotrophs alone runs. Using a computer with a 1 GHz Intel Pentium IV processor, with 500 MB ROM, these simulations took about six hours each.

5.3.3 Competition Between PAOs and Heterotrophs

The researchers tested the competition between heterotrophs and PAOs in ASM2 using both a single aerobic reactor and an anaerobic/aerobic/anoxic series of reactors in which the aerobic reactor had the same DO as the single reactor (Table 5-3). Without anaerobic/aerobic cyclic conditions, PAOs cannot compete with the other heterotrophs. Heterotrophs dominated in the single reactor while PAOs dominated in the reactor series. This confirms that the model reflects the need for an anaerobic/aerobic cyclic condition for the PAOs to compete.

Table 5-3. Competition Between Heterotrophs and PAOs in ASM2.

Scalar (mg/L)	Single Aerobic Reactor	Single Aerobic Reactor	Anaerobic/Aerobic/Anoxic Reactor Series
Influent P	50	150	150
DO	1.0	1.0	0.0 /1.0 / 0.0
Acetate	0.3	0.3	0.0
P	49.7	150	0.91
X_{Het}	664	675	0.0
X_{PAO}	0.02	0.02	697
X_{PP}	0.01	0.01	149
X_{PHA}	0.0	0.0	12

The competition of PAOs and heterotrophs was also tested in the CFD model with different initial conditions and varied oxygen inputs. Because of the increased complexity of including these two microorganisms together, the simulations took much longer to run. For instance, with DO, X_{PP} and X_{PHA} values at 0.05-1.0 mg/L, the mass tolerance in the command file should be set up to be at least 10^{-12} to ensure good precision and stability. However, the lower the mass tolerance the greater the computing time required. For instance, for an oxygen input of 200% (400 g/m^3-day) of the full-scale plant's operating conditions, PAO concentration only changed from 51.7 to 52.75 mg/L in going from 10,000 to 20,000 iterations, which was still far from final steady state results. It is interesting that, in all the CFD simulations, PAOs and heterotrophs co-existed in the system for a relatively large number of iterations when initial conditions started at a value some distance away from steady state. Thus, to achieve more manageable computation times, it was critical to start with appropriate initial conditions.. However, it was encouraging that all simulations for the competition of two organisms were converging, indicating that there was a steady state solution. In some other CFD applications, no steady state exists; in that case simulations have to be run in transient flow mode, which takes much more computing time.

The researchers' strategy was to set initial conditions at near steady-state values based on information obtained from ASM2 and previous CFD simulations with varied oxygen input.

Prolonged simulations were then run to confirm that the concentrations did not diverge from these conditions. Additional simulations were run in which either the PAOs or the heterotrophs were initially input as dominant when they were not expected to be, to ensure that the model calculations would move away from this condition toward the expected steady state. This approach accelerated the testing of the hypothesis that heterotrophs would dominate under high oxygen conditions, and PAOs under low oxygen input conditions, and that heterotrophs and PAOs would coexist in a heterogeneous environment at intermediate oxygen inputs, as can occur in the full-scale plants studied. It was also found that there was numeric instability at initial conditions for polyphosphate mass (X_{PP}) and PHA mass (X_{PHA}) below 2 mg/L. After a great deal of trial and error, the researchers established a range of initial conditions of 5 to 10 mg/L for X_{PP} and X_{PHA}.

The CFD results are summarized in Figure 5-4 and show that, when oxygen is high, the general heterotrophs become dominant and PAOs wash out. When oxygen is too low, both PAO and heterotroph biomass drops. Eventually all biomass is washed out, acetate builds up, and oxygen demand becomes very low because little removal is occurring. This is as expected, since the modeled microorganisms all require aerobic conditions to grow. Again, these results help confirm that the CFD simulations correctly incorporated the biochemical reactions in the fluid field. Of particular interest, however, is the region in which the heterotrophs and PAOs can co-exist, despite the absence of a formal anaerobic zone. The optimum oxygen input for the PAOs is 180 g/m^3-day, which is 90% of the plant operation conditions simulated by ASM1 for nitrogen removal (Daigger and Littleton, 2000). In this simulation, 95% of the tank volume was calculated to have a DO concentration of 0.05-0.3 mg/L.

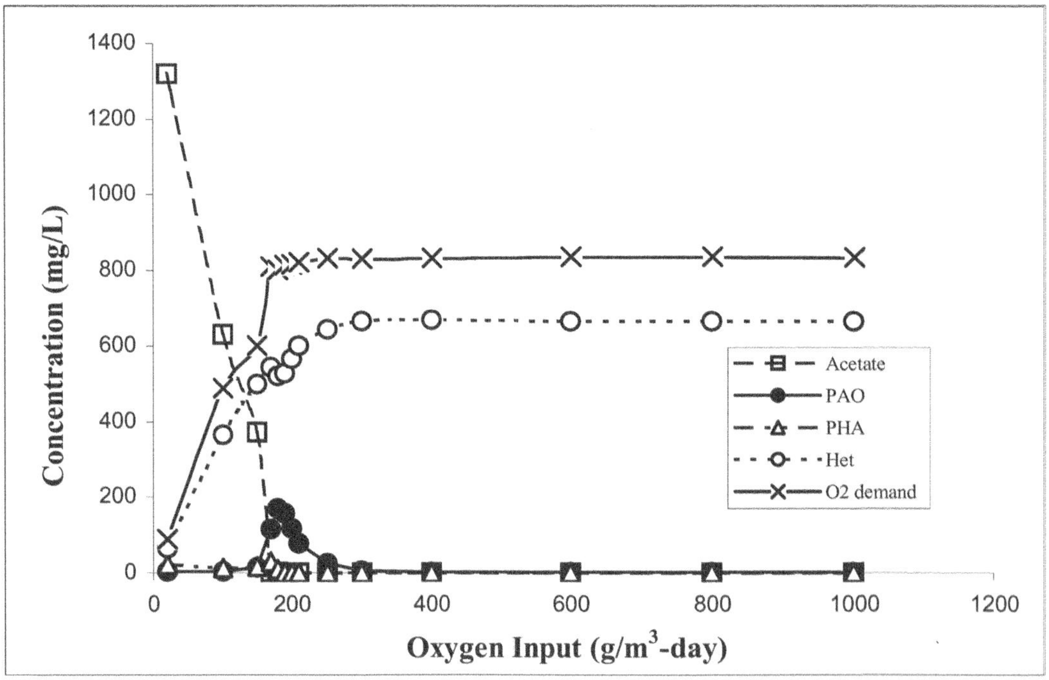

Figure 5-4. PAO *vs.* (AB notes: vs. in Italics) Heterotroph Competition in CFD with Varied Oxygen Input.

Figure 5-5 represents PAO and heterotroph competition in a defined anaerobic and aerobic cyclic condition in ASM2. The competition is controlled by the aerobic volume, which is the percentage of the total bioreactor volume that is aerated. Theoretically, when the aerobic volume is less than 26.3% of the total volume, PAOs will wash out, while if it is less than 4%, the heterotrophs will wash out as well. PAOs are dominant in the region of 40-60% aerobic volume.

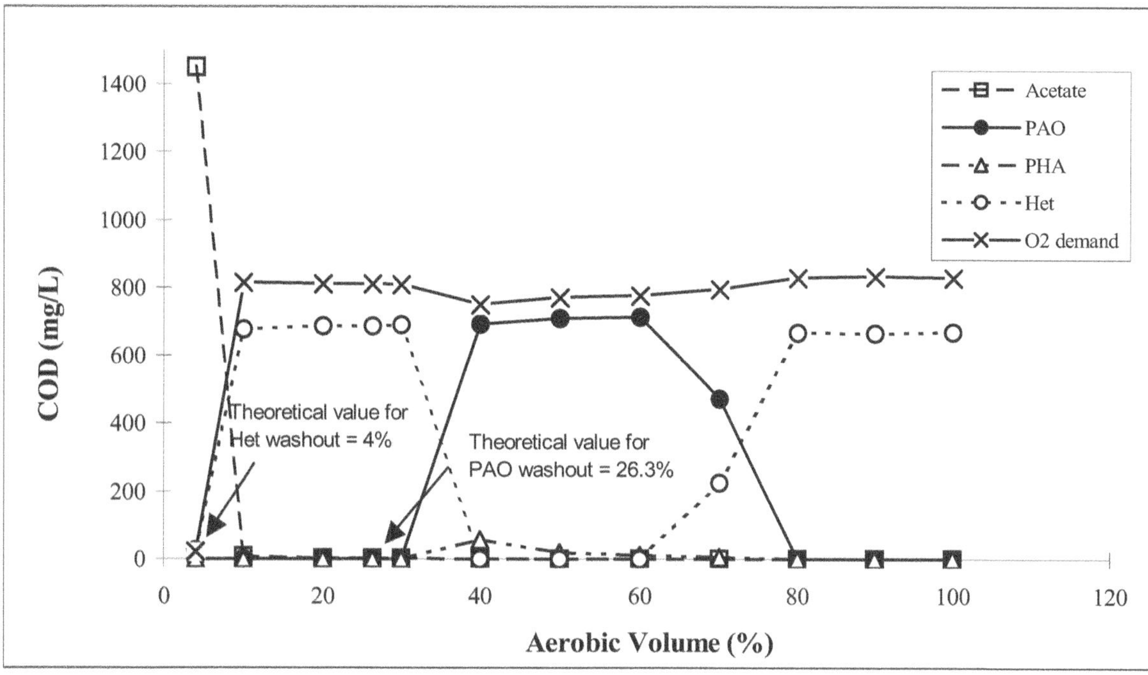

Figure 5-5. PAO *vs.* Heterotroph Competition in ASM2 with Varied Aerobic Volume.

The results for PAO and heterotroph competition in the CFD environment cannot be compared directly to those for the ASM2 simulations because the Orbal^{TM} oxidation ditch is a single bioreactor with undefined anaerobic and aerobic volumes. Nevertheless, oxygen input can be controlled to see how that affects the PAO and heterotroph competition. This gives a similar pattern to the effect of aerobic volume in ASM2.

Figure 5-6 shows the distributions of DO throughout the tank volume for various oxygen inputs in the CFD simulations are shown in. It is clear that a small increase in oxygen input can dramatically affect the DO distribution within the tank. For example, with an oxygen input of 250 g/m^3-day (125% of the original DO input from ASM1), 90% of the tank volume had DO concentrations of 2.5-3.0 mg/L. On the other hand, with 180 g/m^3-day (90% of the original) oxygen input, over 90% of the tank volume had a DO of 0.3 mg/L or less. With 20 g/m^3-day oxygen input (not shown), 99% of the tank volume had a DO below 0.05 mg/L. Figure 5-7 shows the oxygen gradients in the CFD simulation for the optimum oxygen input (180 g/m^3-day) for PAOs in competition with heterotrophs.

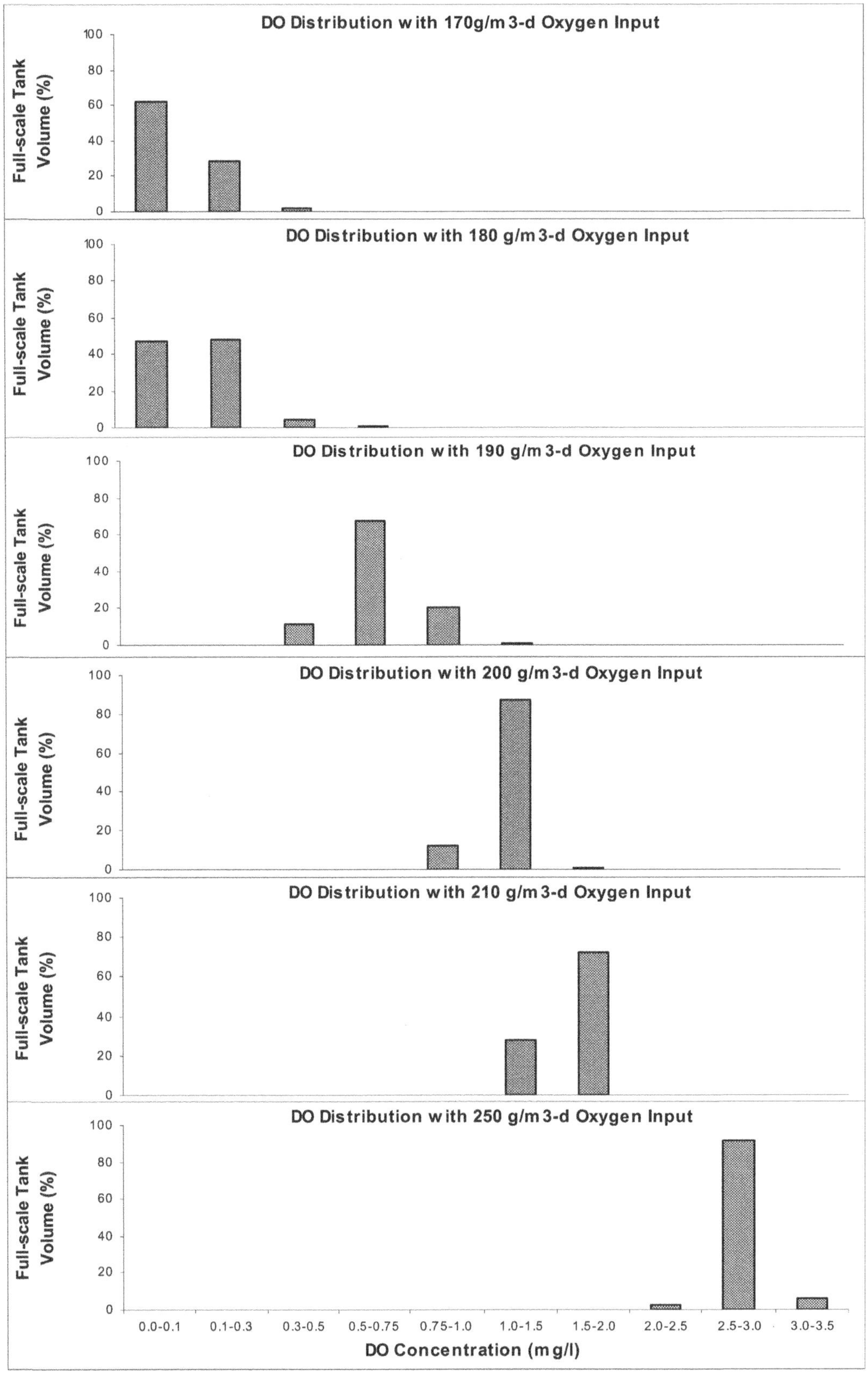

Figure 5-6. DO Distributions in CFD with PAOs and Heterotrophs at Various Oxygen Inputs. Note that the DO concentration ranges are narrower at the lower DO concentrations. (AB notes: this comment isn't mentioned in text.)

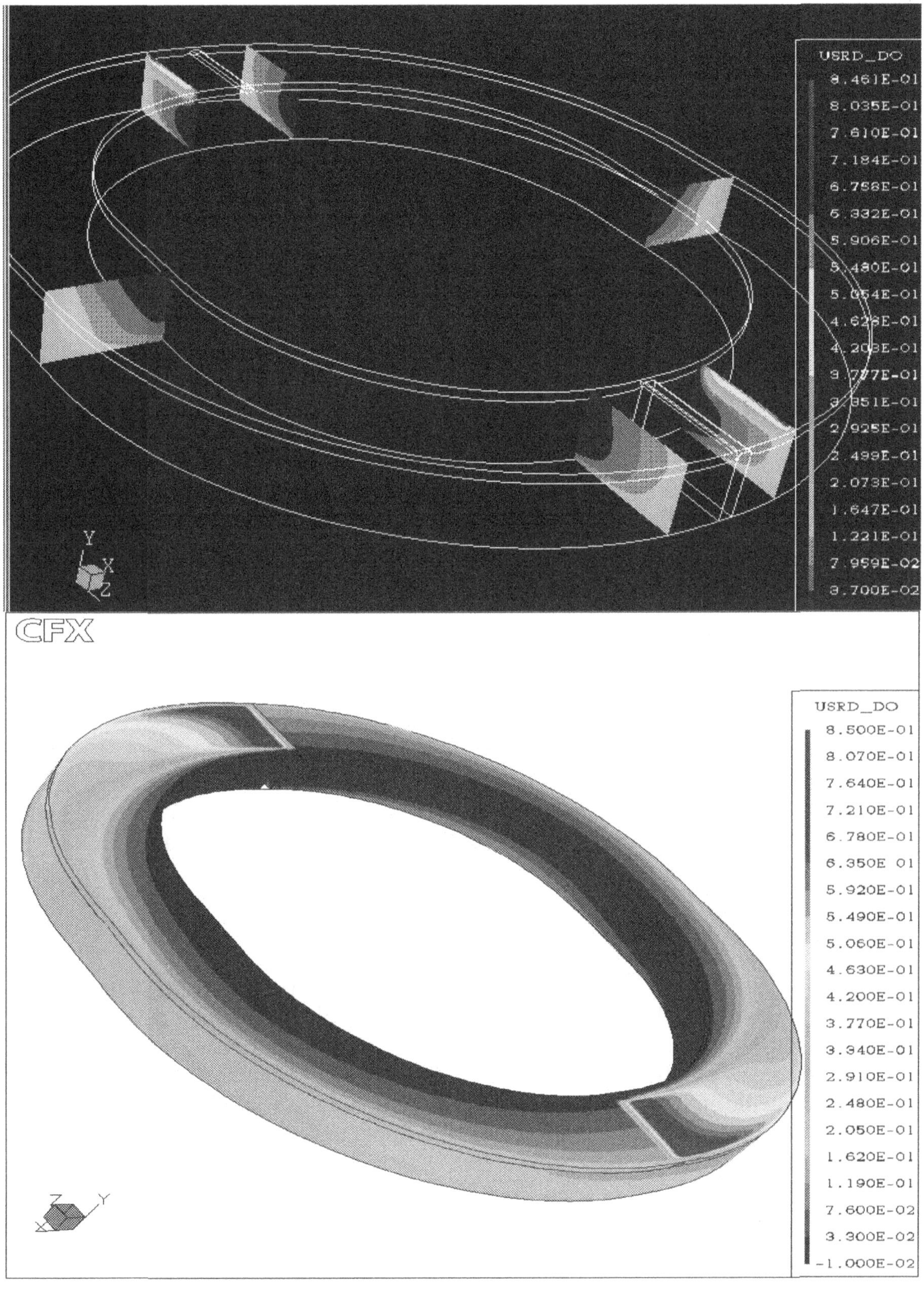

Figure 5-7. DO Gradients from Adding Simplified ASM2 Equations as Source and Sink Terms to CFD Model. (Color figure available on WERF website.)

The CFD results indicate that oxygen input is an important control factor even though Orbal^{TM} plants often operate without detected DO in the outer channel. PAOs and heterotrophs can coexist in the bioreactor when the oxygen input is appropriate to allow sufficient low-DO zones to exist while still providing enough aerobic zones to allow aerobic growth. This finding helps to explain how a small change in DO input can affect nitrification/denitrification and phosphorus removal substantially, leading to a perception that SBNR may be unstable.

5.4 Summary and Conclusions

A self-consistent three-dimensional computational fluid dynamics (CFD) activated sludge model was developed for a simplified full-scale closed-loop bioreactor. Based on ASM2, biochemistry rate expressions for general heterotrophs and phosphorus accumulating organisms (PAOs) were introduced into the previously developed CFD model that characterized the mixing pattern within the outer channel of such a bioreactor. The chemical species of interest were introduced into the CFD environment via source and sink scalar functions in a FORTRAN subroutine to satisfy the Navier-Stokes series of governing equations for conservation of mass and momentum. A simplified approach was used incorporating acetate as the sole carbon and energy source, and no return sludge. Simulations for general heterotrophs and PAOs individually as well in competition were conducted separately in ASM2 and CFD models with varied oxygen input. The following conclusions arise from of this research.

1. The present model provides a useful representation of the transport and biochemical reactions of microorganisms in a closed-loop bioreactor.
2. Both general heterotrophs and PAOs can grow individually in a simplified aerobic closed-loop bioreactor in CFD simulations. These results are consistent with those of ASM2 and an analytical solution.
3. In the CFD model, unlike in ASM2, PAOs and heterotrophs could coexist in a single tank over a range of oxygen inputs. This was possible because there was a second "spiral" current along with the main recirculating flow, creating a varied DO environment.
4. The CFD results demonstrate that oxygen input is an important control factor even when DO concentrations are below normal detection limits.

The investigation reported here is a step towards the development of a mathematical model for realistic closed-loop bioreactors. Further research might extend the model to account for return sludge, nitrification and denitrification, and possibly the multiphase (aqueous and suspended solids) effects of diffusion limitation within the floc. Conceptually, CFD appears capable of predicting the outcomes of the process tank fluid flow pattern and biochemical reactions within the system, but there is still a practical limitation regarding computing time and power.

CHAPTER 6.0

CONCLUSIONS AND RECOMMENDATIONS

Simultaneous biological nutrient removal (SBNR) is the removal of nitrogen and/or phosphorus in excess of that required for biomass synthesis in biological wastewater treatment systems where defined anaerobic and/or anoxic zones do not exist. The basic hypothesis to explain this phenomenon is that three principal mechanisms might be responsible: variations in the bioreactor macroenvironment created by the mixing pattern, gradients within the floc microenvironment, and the activity of novel microorganisms. Within a given system, one or more of these mechanisms might play a major role in SBNR, while others might be of minor importance. Understanding the mechanisms of SBNR can be expected to lead to applying it more efficiently and reliably..

Preliminary work (described in Section 2.2) documented the extent and characteristics of SBNR in seven full-scale Orbal^{TM} closed-loop bioreactors. A relatively simple batch assay was then developed and used to determine the importance of novel microorganisms in SBNR at three of these plants (Section 2.3).

The potential of the floc microenvironment to influence biochemical reactions has long been recognized. Its potential contribution to SBNR was evaluated (Section 2.2) using the IWA's Activated Sludge Model No. 1 (ASM1).

The research described in this report focused on elucidating the role of the bioreactor macroenvironment in SBNR, especially for phosphorus removal. For this purpose, a computational fluid dynamics (CFD) model was developed for a test tank and for the outer channel of a full-scale Orbal^{TM} closed-loop bioreactor. Velocity profiles were determined and compared against field measurements. Biochemical reactions based on IWA's Activated Sludge Model No. 2 (ASM2) were then added for the full-scale system CFD model, and compared in various scenarios to ASM2 predictions for a completely mixed tank or series of tanks, and in some cases to an analytical solution and/or field data.

6.1 Conclusions

Substantial SBNR was observed routinely in full-scale staged, closed-loop bioreactors during the preliminary field study despite the absence of purposely incorporated formal anaerobic or anoxic zones. Essentially complete nitrification occurred, as well as extensive total nitrogen removal . The researchers observed biological phosphorus removal in some of the facilities, as evidenced 1) by a phosphorus-to-BOD_5 removal ratio exceeding that associated simply with biomass synthesis, and 2) by the presence of microorganisms with polyphosphate granules in the mixed liquor.

Nutrient removal was accomplished in these systems by restricting the oxygen input of Channels 1 and 2, which resulted in low DO concentrations there. Both nitrification and denitrification occurred in these channels, resulting in substantial nitrogen removal. The organisms responsible for biological phosphorus removal were also selected for by this operating mode. The extent of nutrient removal appeared to be controlled largely by the oxygen input to these channels—the less oxygen, within limits, the greater the nutrient removal.

ASM1 was used to show that the floc microenvironment mechanism could explain the observed nitrogen removal in these systems. However, this mechanism could not be used to explain the phosphorus removal. It appeared possible that there might have been some vertical stratification (the bioreactor macroenvironment mechanism) of soluble constituents (not mixed liquor suspended solids), resulting in the development of anaerobic zones in the lower portion of Channel 1 (and possibly Channel 2). The activity of novel microorganisms was not evident during this study, and was not necessary to explain the nutrient removal.

The researchers developed a series of batch assays based on the known biochemistry of the novel microorganisms considered, and these assays proved a useful screening tool for the activity of 1) autotrophic denitrification under anoxic conditions and 2) heterotrophic nitrification under aerobic conditions in mixed liquor from full-scale bioreactors accomplishing SBNR. To further simplify the test procedures, the researchers suggested reducing the number of reactors needed (without replication) from four to three for the anoxic test and from four to two for the aerobic test. The results indicated that there was no evidence of substantial activity by novel nitrogen-transforming microorganisms in the three systems tested.

Balanced populations of ammonia- and nitrite-oxidizing bacteria were present in the full-scale mixed liquors tested, as evidenced by the lack of nitrite accumulation during the aerobic batch assays. This suggests that ammonia is being converted to nitrate in the full-scale systems before it is subsequently denitrified. One set of batch test results with a very low aeration rate also demonstrated that nitrification and denitrification could occur simultaneously without detected DO.

Anaerobic/aerobic batch assays assessed the activity of PAOs at one plant. Anaerobic phosphate release (following acetate addition) and aerobic phosphate uptake were observed as expected, based on current understanding of PAO metabolism. The zero-order phosphate uptake rate suggested that PAOs composed ~13% of the process mixed liquor. These results, coupled with previous observations of process phosphate removal, mixed liquor phosphate content, and the presence of intracellular polyphosphate granules, strongly suggest that excess biological phosphorus removal was occurring in these systems. These results also demonstrate that the mixed liquor was capable of phosphate release, even though such release was not observed in the full-scale system, and no evidence of novel microorganisms was found.

The CFD models developed for the test tank and for the outer channel of a closed-loop oxidation ditch of the Orbal™ configuration used a simplified approach to create fluid motion in the two tanks. This consisted of imparting a fraction of the tank's observed momentum (based on measured velocity) uniformly across the location of the disc aerators in a horizontal direction. The flow patterns predicted for each tank yielded a good degree of fit with the measured velocities. Small systematic errors were observed, probably due to the simplifications employed, but overall the CFD model captured well the mixing pattern generated within the tanks. This complex flow pattern consisted of a high velocity discharge from the aerator that flowed downward to the bottom of the tank at the first curve and then upward to the top of the tank on the side opposite of the aerator. The resulting flow thus had a spiral form.

The researchers evaluated the potential for DO variations in the full-scale tank by including a source of oxygen at the location of the disc aerators and an oxygen sink distributed throughout the contents of the full-scale bioreactor. Even with this highly simplified form for adding biochemical equations to the CFD model, zones of relatively higher and lower DO concentrations occurred in the bioreactor simulation. These results suggested that the heterogeneous flow pattern created in the bioreactor macroenvironment could create informal anaerobic and anoxic zones that would allow SBNR to occur in such aerated systems.

The researchers then introduced more realistic, though still simplified, biochemical rate expressions based on ASM2 for general heterotrophs and phosphorus accumulating organisms (PAOs) into the previously developed CFD model that characterized the mixing pattern within the outer channel of the full-scale bioreactor. The chemical species of interest were introduced into the CFD model via source and sink scalar functions in a FORTRAN subroutine to satisfy the Navier-Stokes series of governing equations for conservation of mass and momentum. The simulations were performed using acetate as the sole carbon and energy source, and with no return sludge. Separately, in ASM2 and CFD models with varied oxygen input, the researchers conducted simulations for general heterotrophs and PAOs individually and for the two organisms in competition.

The CFD model provided useful simulations to qualitatively describe the bioreactor system. This allowed testing of the hypothesis that the bioreactor macroenvironment contributes to SBNR, especially with respect to phosphorus. CFD handled simulations of heterotrophs or PAOs alone in a reasonable computational period. The model still converged on a solution when the equations for both organisms were included, so that they competed against each other. However, under these conditions the computations took days to complete when near steady state values were provided as the initial conditions; when initial conditions were not near steady state, a solution was not complete even after weeks of computing time.

The CFD model predicted that either heterotrophs or PAOs alone could grow in the simulated reactor. These results were consistent with those predicted by ASM2, and, for heterotrophs, by an analytical solution.

However, unlike the ASM2 results, the CFD model shows that heterotrophs and PAOs could coexist in a single tank over a range of oxygen inputs. This was possible because there was a second "spiral" current along with the main recirculating flow, creating a varied DO environment. The CFD results also demonstrated that oxygen input to a reactor is an important control factor even when DO concentrations are below normal detection limits.

This work is a step towards the development of a mathematical model for realistic closed loop bioreactors. Conceptually, CFD appears capable of predicting the outcomes of the process tank fluid flow pattern and biochemical reactions within realistic systems, but there is still a practical limitation regarding model complexity and computing resources.

6.2 Recommendations

This project represents the first application of computational fluid dynamics to the activated sludge biological treatment system. CFD was used to show that the mixing patterns in a closed-loop tank bioreactor macroenvironment could unintentionally create the anaerobic and anoxic zones necessary for simultaneous biological nutrient removal.

CFD has the potential to provide more quantitative information about real systems. However, its usefulness in this regard is limited at present by the model complexity, cost, and computational requirements. For now, these problems may limit CFD to research applications. As the software and computing systems improve, however, more regular use of CFD will become feasible. This may include opportunities to use CFD for troubleshooting at existing facilities as well as for designing wastewater treatment bioreactors.

Incremental steps that can be taken to enhance the CFD model include incorporation of return sludge, nitrification and denitrification reactions, the inner channels, and perhaps the effects of diffusion limitations within the biological floc. It should also be possible to refine the method of incorporating the momentum imparted by the aerators.

One exciting prospect that should be explored in future work is the possibility of designing or operating bioreactors to intentionally control DO gradients within a tank to optimize treatment. (Although low DO bulking does not seem to be a problem with this approach, it would still need to be considered in any full-scale application.) CFD can potentially play a major role in this effort. Variables may include tank shape and dimensions, and aerator rpm and submergence. Another area to explore would be incorporating simple baffles to promote spiral flow or otherwise increase the DO gradients created. This has the potential to very economically enhance nutrient removal in existing facilities.

REFERENCES

Albertson, O.E. and H.D. Stensel (1994) "Aerated Anoxic Biological NdeN Process," *Water Sci. Technol.*, **29(7)**, 167-176.

Albertson, O.E. and J. Coughenour (1995) "Aerated Anoxic Oxidation-Denitrification Process," *J. Environ. Eng.*, **121**, 720-726.

Anderson, J.D., Jr. (1995) *Computational Fluid Dynamics*, McGraw-Hill, Inc., New York.

Anon. (1997) "CFX4 Solver," AEA (now ANSYS) Technology, Pittsburgh, PA.

Anon. (2000) *CFX Manual*, AEA (now ANSYS) Technology, Pittsburgh, PA.

Anon. (2002) CFX User Conference, May 8-9, Pittsburgh, PA.

Applegate, C.S., B. Wilder, and J.R. DeShaw (1980) "Total Nitrogen Removal in a Multi-Channel Oxidation System," *J. Water Pollut. Control Fed.*, **52**, 568-577.

Barker, P.S. and P.L. Dold (1997a) "General Model for Biological Nutrient Removal Activated-Sludge Systems: Model Presentation," *Water Environ. Res.*, **69**, 969-984.

Barker, P.S. and P.L. Dold (1997b) "General Model for Biological Nutrient Removal Activated-Sludge Systems: Model Application," *Water Environ. Res.*, **69**, 985-991.

Bistrup, S.M. and C.P.L. Grady, Jr. (1988) "SSSP – Simulation of Single-Sludge Processes," *J. Water Pollut. Control Fed.*, **60**, 351-361.

Bock, E., I. Schmidt, R. Stuven, and D. Zart (1995) "Nitrogen Loss Caused by Denitrifying *Nitrosomonas* Cells Using Ammonium or Hydrogen as Electron Donors and Nitrite as Electron Acceptor," *Arch. Microbiol.*, **163**, 16-20.

Brewer, H., J.P. Stephenson, and D. Green (1995) "Plant Optimization Using Online Phosphorus Analyzers and Automated SRT Control to Achieve Harbour Delisting," *Proceedings of Water Environment Federation 68th Annual Conference & Exposition, Volume 5, Management and Facility Operations*, 673-684.

Cinar, O., G.T. Daigger, and S.P. Graef (1998) "Evaluation of IAWQ Activated Sludge Model No. 2 Using Steady-State Data from Four Full-Scale Wastewater Treatment Plants," *Water Environ. Res.*, **70**, 1216-1224.

Clesceri, L.S., A.E. Greenberg, A.D. Eaton, Eds. (1998) *Standard Methods for the Examination of Water and Wastewater, 20th Edition*, American Public Health Association, American Water Works Association, and Water Environment Federation, Washington, D.C.

Daigger, G.T. and H.X. Littleton (1999) "Characterization of Simultaneous Nutrient Removal in Staged, Closed Loop Bioreactors," *Proceedings of the Water Environment Federation 72nd Annual Conference & Expositon*, New Orleans, LA [CD-ROM].

Daigger, G.T. and H.X. Littleton (2000) "Characterization of Simultaneous Nutrient Removal in Staged, Closed Loop Bioreactors," *Water Environ. Res.*, **72**, 330-339.

Daigger, G.T., G.V. Crawford, M. O'Shaughnessey, and M. D. Elliott (1998) "The Use of Coupled Refined Stoichiometric and Kinetic/Stoichiometric Models to Characterize Entire Wastewater Treatment Plants," *Proceedings, Water Environment Federation 71st Annual Conference & Exposition, Volume 1, Wastewater Treatment Research and Municipal Wastewater Treatment*, Orlando, FL, 617-628.

Dobbeck, P. (1999) US Filter/Envirex, "Orbal Channel Computational Fluid Dynamic Analysis," Internal Report.

Dobbeck, P., G. Smith, C. Applegate, and P. Petit (1999) US Filter/Envirex, Personal communication.

Draper, N.R. and H. Smith (1966) *Applied Regression Analysis*, John Wiley & Sons, Inc., New York.

Drews, R.J.L.C. and A.M. Greef (1973) "Nitrogen Elimination by Rapid Alternation of Aerobic/Anoxic Conditions in Orbal Activated Sludge Plants," *Water Res.*, **7**, 1183-1194.

Ducoste, J. and G.T. Daigger (1999) "Evaluation of Stacked Secondary Clarifier Design Using Computational Fluid Dynamics, *Proceedings of the Water Environment Federation 72nd Annual Conference & Exposition*, New Orleans, LA [CD-ROM].

Filipe, C.D.M. (1999) McMaster University, Personal Communication, June.

Filipe, C.D.M. and G.T. Daigger (1997) "Evaluation of the Capacity of Phosphorus-Accumulating Organisms to Use Nitrate as Well as Oxygen as Final Electron Acceptor: A Theoretical Study on Population Dynamics," *Proceedings, Water Environment Federation 70th Annual Conference & Exposition, Volume 1, Research and Municipal Wastewater Treatment*, Dallas, TX, 341-352.

Filipe, C.D.M. and G.T. Daigger (1998) "Development of a Revised Metabolic Model for the Growth of Phosphorus Accumulating Organisms," *Water Environ. Res.*, **70**, 670.

Filipe, C.D.M. and G.T. Daigger (1999) "Evaluation of the Capacity of Phosphorus-Accumulating Organisms to Use Nitrate and Oxygen as Final Electron Acceptors: A Theoretical Study on Population Dynamics," *Water Environ. Res.*, **71**, 1140-1150.

Fletcher, C.A.J. (1991) *Computational Techniques for Fluid Dynamics Volume 1: Fundamentals and General Techniques*, Springer-Verlag, Berlin, Germany.

Geraats, S.G.M., C.M. Hooijmans, E.W.J. van Niel, L.A. Robertson, J.J. Heijnen, K.Ch.A.M Lyben, and J.G. Kuenen (1990) "The Use of a Metabolically Structured Model in the Study of Growth, Nitrification, and Denitrification by *Thiosphaera pantotropha*," *Biotechnol. Bioeng.*, **36**, 921-930.

Grady, C.P.L., Jr., G.T. Daigger, and H.C. Lim (1999) *Biological Wastewater Treatment*, 2nd Edition, Revised and Expanded, Marcel Dekker, New York.

Henze, M., C.P.L. Grady, Jr., W. Gujer, G.v.R. Marais, and T. Matsuo (1987) "Model for Single-Sludge Wastewater Treatment," *Water Res.*, **21**, 505.

Henze. M., W. Gujer, T. Mino, T. Matsuo, M.C. Wentzel, and G.v.R. Marais (1995) "Activated Sludge Model No. 2," *IAWQ Scientific and Technical Reports, No.3.*

Henze M., W. Gujer, T. Mino, T. Matsuo, M. Wentzel, G.v.R.Marais, and M. Van Loosdrecht (1999) "Activated Sludge Model No. 2(AB asks: delete d?) by IAWQ Task Group on Mathematical Modelling for Design and Operation of Biological Wastewater Treatment Processes". *Water Sci. Technol.*, **39** (1), 165-182.

Huang D-H, J. Peccia, J.L. Zilles, and D.R. Noguera (2002) "Physical Enrichment of Polyphosphate-Accumulating Organisms in Activated Sludge," *Water Environ. Res.*, **74**, 354-361.

International Water Association (2000) "Activated Sludge Models ASM1, ASM2, ASM2d and ASM3," *Scientific and Technical Report No. 9*, International Water Association, London.

Jenkins, D. (1999) Univ. of California at Berkeley, Personal Communication, April.

Jetten, M.S.M., M. Strous, K.T. van de Pas-Schoonen, J. Schalk, U.G.J.M. van Dongen, A.A. van de Graaf, S. Logemann, G. Muyzer, M.C.M. van Loosdrecht, and J.G. Kuenen (1999) "The Anaerobic Oxidation of Ammonium," *FEMS Microb. Rev.*, **22**, 421-437.

Jetten M.S.M., M. Schmid, I. Schmidt., M. Wubben, U. van Dongen, W. Abma, O. Sliekers, N.P. Revsbech, H.J.E. Beaumont, L. Ottosen, E. Volcke, H.J. Laanbroek, J.L. Campos-Gomez, J. Cole, M.C.M. van Loosdrecht, J.W. Mulder, J. Fuerst, D. Richardson, K. van de Pas, R. Mendez-Pampin, K. Third, I. Cirpus, R. van Spanning, A. Bollmann, L.P. Nielsen, H.O. den Camp, C. Schultz, J. Gundersen, P. Vanrolleghem, M. Strous, M. Wagner, and J.G. Kuenen (2002) "Improved Nitrogen Removal by Application of New Nitrogen-cycle Bacteria," *Re/Views in Environmental Science & Bio/Technology*, **1,** 51-63.

Kuai, L. and W. Verstraete (1998) "Ammonium Removal by the Oxygen-Limited Autotrophic Nitrification-Denitrification System," *Appl. Environ. Microbiol.*, **64**, 4500-4506.

Kuba, T., M.C.M. van Loosdrecht, F.A. Brandse, and J.J. Heijnen (1997) "Occurrence of Denitrifying Phosphorus Removing Bacteria in Modified UCT-Type Wastewater Treatment Plants," *Water Res.*, **31**, 777-786.

Lau, A.O., P.F. Strom, and D. Jenkins (1984) "The Competitive Growth of Flocforming and Filamentous Bacteria: A Model for Activated Sludge Bulking," *J. Water Pollut. Control Fed.*, **56**, 52-61.

Lesieur, M., P. Comte, and Z. Justin (1996) *Computational Fluid Dynamics*, North-Holland Elsevier, New York.

Li, B. and P. Bishop (2001) "ORP Changes in Aerations Tanks and Microprofiles of Activated Sludge Floc in the Medium and Low Pollutant Strength Wastewaters," *Proceedings of the Water Environment Federation 74th Annual Conference & Exposition on Water Quality and Wastewater Treatment*, Atlanta, GA [CD-ROM].

Li, B. and P. Bishop (2003) "Structure-Function Dynamics and Modeling Analysis of the Micro-Environment of Activated Sludge Floc" *Water Sci. Technol.*, **47** (11), 267-273.

Littleton, H.X., G.T. Daigger, P.F. Strom, and R.A. Cowan (2000) "Evaluation of Autotrophic Denitrification and Heterotrophic Nitrification in Simultaneous Biological Nutrient Removal Systems," *Proceedings of the Water Environment Federation 73rd Annual Conference & Exposition on Water Quality and Wastewater Treatment*, Anaheim, CA [CD-ROM].

Littleton, H.X., G.T. Daigger, and P.F. Strom (2001) "Application of Computational Fluid Dynamics to Closed Loop Bioreactors–Analysis of Macro-environment. Variations in Simultaneous Biological Nutrient Removal Systems," *Proceedings of the Water Environment Federation 74th Annual Conference & Exposition*, Atlanta, GA [CD-ROM].

Littleton, H.X., G.T. Daigger, P.F. Strom, and R.M. Cowan (2002a) "Evaluation of Autotrophic Denitrification, Heterotrophic Nitrification, and PAOs in Full Scale Simultaneous Biological Nutrient Removal Systems," *Water Sci. Technol.*, **46** (1-2), 305–312.

Littleton, H.X., G.T. Daigger, P.F. Strom, and R. Jin (2002b) "Simulation of BioP Removal in CFD Environment - Analysis of Macro-environment Variations in Simultaneous Biological Nutrient Removal Systems," *Proceedings of the Water Environment Federation 75th Annual Technical Exposition & Conference*, Chicago, IL [CD-ROM].

Littleton, H.X., G.T. Daigger, P.F. Strom, and R.A. Cowan (2003a) "Simultaneous Biological Nutrient Removal: Evaluation of Autotrophic Denitrification, Heterotrophic Nitrification, and Biological Phosphorus Removal in Full-Scale Systems," *Water Environ. Res.*, **75**, 138-150.

Littleton, H.X., G.T. Daigger, and P.F. Strom (2003b) "Summary Paper: Mechanisms of Simultaneous Biological Nutrient Removal in Closed Loop Bioreactors," *Proceedings of the Water Environment Federation 76th Annual Conference & Exposition on Water Quality and Wastewater Treatment*, Los Angeles, CA [CD-ROM].

Madigan, M.T., J.M. Martinko, and J. Parker (2003) *Brock Biology of Microorganisms, 10th Edition*, Prentice Hall, Upper Saddle River, NJ.

Metcalf & Eddy, Inc. (1991) *Wastewater Engineering Treatment/ Proposal/Reuse, 3rd Edition*, McGraw-Hill, Inc., New York.

Mino, T., M.C.M. van Loosdrecht, and J.J. Heijnen (1998) "Review Paper: Microbiology and Biochemistry of the Enhanced Biological Phosphate Removal Process," *Water Res.*, **32**, 3193-3207.

Noguera, D.R., C.-H. Huang, and J.M. Regan (2000) "Separation of Polyphosphate-Accumulating Bacteria from Activated Sludge Samples Using Density Gradient Centrifugation and Flow Cytometry," *Proceedings of the Water Environment Federation 73rd Annual Conference & Exposition on Water Quality and Wastewater Treatment*, Anaheim, CA [CD-ROM].

Ostle, B. (1963) *Statistics in Research*, The Iowa State University Press, Ames, IA.

Park, H-D, J.M. Regan, and D.R. Noguera (2001) "Molecular Analysis of Ammonia-Oxidizing Bacterial Populations in Aerated-Anoxic Orbal Processes," *Proceedings of the 3rd IWA International Specialised Conference Microorganisms in Activated Sludge and Biofilm Processes*, Rome, 202-209.

Park, H-D, J.M. Regan, and D.R. Noguera (2002) "Molecular Analysis of Ammonia-Oxidizing Bacterial Populations in Aerated-Anoxic Orbal Processes," *Water Sci. Technol.*, **46**, 273-280.

Randall, C.W., J.L. Barnard, and H.D. Stensel, Editors (1992) *Design and Retrofit of Wastewater Treatment Plants for Biological Nutrient Removal*, Technomics Publishing, Lancaster, PA.

Rittman, B.E. and W.E. Langeland (1985) "Simultaneous Denitrification with Nitrification in Single-Channel Oxidation Ditches," *J Water Pollut. Control Fed.*, **57**, 300-308.

Robertson, L.A. and J.G. Kuenen (1984) "Aerobic Denitrification –Old Wine in New Bottles?," *Antonie van Leeuwenhoek*, **50**, 525-544.

Robertson, L.A. and J.G. Kuenen (1990) "Combined Heterotrophic Nitrification and Aerobic Denitrification in *Thiosphaera pantotropha* and Other Bacteria," *Antonie van Leeuwenhoek*, **56**, 289-299.

Robertson, L.A., E.W.J. van Niel, R.A.M. Torremans, and J.G. Kuenen (1988) "Simultaneous Nitrification and denitrification in Aerobic Chemostat Cultures of *Thiosphaera pantotropha*," *Appl. Environ. Microbiol.*, **54**, 2812-2818.

Smith, G. (1996) "Increasing Oxygen Delivery in Anoxic Tanks to Improve Denitrification," *Proceedings of the Water Environment Federation 69th Annual Conference & Exposition, Volume 5, Management and Innovative and Alternative Technologies*, Chicago, 345-405.

Stenstrom M.K. and S.S. Song (1991) "Effect of Oxygen Transport Limitation on Nitrification in the Activated Sludge Process," *Res. J. Water Pollut. Control Fed.* **63**, 208-219.

Strous, M., E. van Gerven, P. Zheng, J.G. Kuenen, and M.S.M. Jetten (1997) "Ammonium Removal From Concentrated Waste Streams with the Anaerobic Ammonium Oxidation (ANAMMOX) process in Different Reactor Configurations," *Water. Res.*, **31**, 1955-1962.

Strous, M., J.G. Kuenen, and M.S.M. Jetten (1999a) "Key Physiology of Anaerobic Ammonium Oxidation," *Appl. Environ. Microbiol.*, **65**, 3248-3250.

Strous, M., J.A. Fuerst, E.H.M. Kramer, S. Logemann, G. Muyzer, K.T. van de Pas-Schoonen, R. Webb, J.G. Juenen, and M.S.M. Jetten (1999b) "Missing Lithotroph Identified as New Planctomycete," *Nature (London)*, **400**, 446-449.

U.S. EPA, (1983) *Methods for Chemical Analysis of Water and Wastes*, USEPA –600/14-79-020; Office of Research and Development, Cincinnati, OH, Method 365.2.

U.S. EPA (1987) *Technology Transfer Design Manual Phosphorus Removal*, USEPA/625/1-87/001 CER1, Cincinnati, OH.

Van der Walt, J.J. (2000) "Reflecting on CFD efforts in water industry," *Proceedings of IWA Workshop, Modeling of Conventional Drinking Water Production Processes*, Delft, The Netherlands.

Van Huyssteen, J.A., J.L. Barnard, and J. Hendriksz (1990) "The Olifantsfontein Nutrient Removal Plant," *Water Sci. Technol.*, **22(7/8)**, 1-8.

Van Loosdrecht, M.C.M. and M.S.M. Jetten (1998) "Microbiological Conversions in Nitrogen Removal," *IAWQ 19th Biennial International Conference Preprint Book 1, Nutrient Removal*, 1-8.

van Munch, E., P. Land, and J. Keller (1996) "Simultaneous Nitrification and Denitrification in Bench-Scale Sequencing Batch Reactors," *Water Sci. Technol.*, **30(2)**, 277-284.

Van Niekirk, A., D. Jenkins, and M.G. Richards (1988) "A Mathematical Model of the Carbon-Limited Growth of Filamentous and Floc-Forming Organisms in Low F/M Sludge," *J. Water Pollut. Control Fed.*, **60**, 100-106.

Versteeg, H.K. and W. Malalasekera (1995) *An Introduction to Computational Fluid Dynamics*, Prentice Hall, Upper Saddle River, NJ.

Warsi, Z.U.A. (1993) *Fluid Dynamics: Theoretical and Computational Approaches*, CRC Press Inc., Boca Raton, FL.

Wei, T., E.M. Kline, S.H.-K. Lee, and S. Woodruff (1992) "Gortler Vortex Formation at the Inner Cylinder in Taylor-Couette Flow," *J. Fluid Mech.*, **245**, *pp*.47-68.

Zilles, J.L., C-H Hung, and D.R. Noguera (2001) "Presence of *Rhodocyclus* in Full-Scale Wastewater Treatment Plant and Their Participation in Enhanced Biological Phosphorus Removal," *Proceedings of the 3rd IWA International Specialised Conference Microorganisms in Activated Sludge and Biofilm Processes*, Rome, 75-81.

Zilles, J.L., J. Peccia, and D.R. Noguera (2002) "Microbiology of Enhanced Biological Phosphorus Removal in Aerated-Anoxic Orbal Processes," *Water Environ. Res.*, **74**, 428-436.

WASTEWATER UTILITY

Alabama
Montgomery Water Works & Sanitary Sewer Board

Alaska
Anchorage Water & Wastewater Utility

Arizona
Gila Resources

Glendale, City of, Utilities Department

Mesa, City of

Peoria, City of

Pima County Wastewater Management

Arkansas
Little Rock Wastewater Utility

California
Calaveras County Water District

Central Contra Costa Sanitary District

Contra Costa Water District

Crestline Sanitation District

Delta Diablo Sanitation District

Dublin San Ramon Services District

East Bay Dischargers Authority

East Bay Municipal Utility District

El Dorado Irrigation District

Fairfield-Suisun Sewer District

Fresno Department of Public Utilities

Irvine Ranch Water District

Las Virgenes Municipal Water District

Livermore, City of

Lodi, City of

Los Angeles, City of

Los Angeles County, Sanitation Districts of

Napa Sanitation District

Orange County Sanitation District

Palo Alto, City of

Riverside, City of

Sacramento Regional County Sanitation District

San Diego Metropolitan Wastewater Department, City of

San Francisco, City & County of

San Jose, City of

Santa Barbara, City of

Santa Rosa, City of

South Bayside System Authority

South Coast Water District

South Orange County Wastewater Authority

Sunnyvale, City of

Union Sanitary District

West Valley Sanitation District

Colorado
Boulder, City of

Colorado Springs Utilities

Littleton/Englewood Water Pollution Control Plant

Metro Wastewater Reclamation District, Denver

Connecticut
The Mattabasset District

New Haven, City of, WPCA

District of Columbia
District of Columbia Water & Sewer Authority

Florida
Broward, County of

Fort Lauderdale, City of

Gainesville Regional Utilities

JEA

Miami-Dade Water & Sewer Authority

Orange County Utilities Department

Orlando, City of

Reedy Creek Improvement District

Seminole County Environmental Services

St. Petersburg, City of

Stuart Public Utilities

Tallahassee, City of

Tampa, City of

Toho Water Authority

West Palm Beach, City of

Georgia
Atlanta Department of Watershed Management

Augusta, City of

Clayton County Water Authority

Cobb County Water System

Columbus Water Works

Fulton County

Gwinnett County Department of Public Utilities

Savannah, City of

Hawaii
Honolulu, City and County of

Illinois
American Bottoms Wastewater Treatment Plant

Greater Peoria Sanitary District

Kankakee River Metropolitan Agency

Metropolitan Water Reclamation District of Greater Chicago

North Shore Sanitary District

Wheaton Sanitary District

Iowa
Ames, City of

Cedar Rapids Wastewater Facility

Des Moines Metro Wastewater Reclamation Authority

Iowa City, City of

Kansas
Johnson County Unified Wastewater Districts

Unified Government of Wyandotte County/Kansas City, City of

Kentucky
Louisville & Jefferson County Metropolitan Sewer District

Louisiana
Sewerage & Water Board of New Orleans

Maine
Bangor, City of

Portland Water District

Maryland
Anne Arundel County Bureau of Utility Operations

Howard County Department of Public Works

Washington Suburban Sanitary Commission

Massachusetts
Boston Water & Sewer Commission

Upper Blackstone Water Pollution Abatement District

Michigan
Ann Arbor, City of

Detroit, City of

Holland Board of Public Works

Lansing, City of

Owosso Mid-Shiawassee County WWTP

Saginaw, City of

Wayne County Department of Environment

Wyoming, City of

Minnesota
Rochester, City of

Western Lake Superior Sanitary District

Missouri
Independence, City of

Kansas City Missouri Water Services Department

Little Blue Valley Sewer District

Nebraska
Lincoln Wastewater System

Nevada
Henderson, City of

New Jersey
Bergen County Utilities Authority

Passaic Valley Sewerage Commissioners

New York
New York City Department of Environmental Protection

Rockland County Solid Waste Management Authority/ Sewer District

North Carolina
Charlotte/Mecklenburg Utilities

Durham, City of

Metropolitan Sewerage District of Buncombe County

Orange Water & Sewer Authority

Ohio
Akron, City of

Butler County Department of Environmental Services

Columbus, City of

Metropolitan Sewer District of Greater Cincinnati

Northeast Ohio Regional Sewer District

Summit, County of

Oklahoma
Tulsa, City of

Oregon
Clean Water Services

Eugene/Springfield Water Pollution Control

Water Environment Services

Pennsylvania
Philadelphia, City of

University Area Joint Authority, State College

South Carolina
Charleston Commissioners of Public Works

Mount Pleasant Waterworks & Sewer Commission

Spartanburg Sanitary Sewer District

Tennessee
Cleveland, City of

Knoxville Utilities Board

Murfreesboro Water & Sewer Department

Nashville Metro Water Services

Texas
Austin, City of

Dallas Water Utilities

Denton, City of

El Paso Water Utilities

Fort Worth, City of

Gulf Coast Waste Disposal Authority

Houston, City of

San Antonio Water System

Trinity River Authority

Utah
Salt Lake City Corporation

Virginia
Alexandria Sanitation Authority
Arlington, County of
Fairfax County Virginia
Hampton Roads Sanitation District
Henrico, County of
Hopewell Regional Wastewater Treatment Facility
Loudoun County Sanitation Authority
Lynchburg Regional WWTP
Prince William County Service Authority
Richmond, City of
Rivanna Water & Sewer Authority

Washington
Edmonds, City of
Everett, City of
King County Department of Natural Resources
Seattle Public Utilities
Sunnyside, Port of
Yakima, City of

Wisconsin
Green Bay Metro Sewerage District
Madison Metropolitan Sewerage District
Milwaukee Metropolitan Sewerage District
Racine, City of
Sheboygan Regional Wastewater Treatment
Wausau Water Works

Australia
South Australian Water Corp.
Sydney Water Corp.
Water Corp. of Western Australia

Canada
Greater Vancouver Regional District
Toronto, City of, Ontario
Winnipeg, City of, Manitoba

Mexico
Servicios de Agua y Drenaje de Monterrey, I.P.D.

New Zealand
Watercare Services Limited

Singapore
Singapore Public Utilities Board

United Kingdom
Yorkshire Water Services Limited

STORMWATER UTILITY

California
Los Angeles, City of, Department of Public Works
Monterey, City of
San Francisco, City & County of
Santa Rosa, City of
Sunnyvale, City of

Colorado
Boulder, City of

Georgia
Griffin, City of

Iowa
Cedar Rapids Wastewater Facility
Des Moines Metro Wastewater Reclamation Authority

Kansas
Overland Park, City of

Kentucky
Louisville & Jefferson County Metropolitan Sewer District

Maine
Portland Water District

Minnesota
Western Lake Superior Sanitary District

North Carolina
Charlotte, City of, Stormwater Services

Pennsylvania
Philadelphia, City of

Tennessee
Chattanooga Storm Water Management

Washington
Seattle Public Utilities

STATE

Arkansas Department of Environmental Quality
Fresno Metropolitan Flood Control District, Calif.
Urban Drainage & Flood Control District, Colo.

CORPORATE

ADS Environmental Services
The ADVENT Group Inc.
Alan Plummer & Associates
Alden Research Laboratory
Aqua-Aerobic Systems, Inc.
Aquateam–Norwegian Water Technology Centre A/S
BaySaver Inc.
BioVir Laboratories, Inc.
Black & Veatch
Boyle Engineering Corporation
Brown & Caldwell
Burns & McDonnell
CABE Associates Inc.
The Cadmus Group
Camp Dresser & McKee Inc.
Carollo Engineers Inc.
Carpenter Environmental Associates Inc.
CDS Technologies Inc.
Chemtrac Systems Inc.
CH2M HILL
Damon S. Williams Associates, LLC
David L. Sheridan, P.C.
Dewling Associates, Inc.
Earth Tech Inc.
Eco-Matrix
Ecovation
Environmental Engineers International
EMA Inc.
The Eshelman Company Inc.
Finkbeiner, Pettis, & Strout Inc.
Freese & Nichols, Inc.
ftn Associates Inc.
Fuss & O'Neill, Inc.
Gannett Fleming Inc.
Golder Associates Inc.
Greeley and Hansen LLC
Hazen & Sawyer, P.C.
HDR Engineering Inc.
HNTB Corporation
HydroQual Inc.
Infilco Degremont Inc.
Ingersoll-Rand Energy Systems
Insituform Technologies Inc.
Institute for Environmental Technology & Industry, Korea
Jacobson Helgoth Consultants Inc.
Jason Consultants Inc.
Jordan, Jones, & Goulding Inc.
KCI Technologies Inc.
Kelly & Weaver, P.C.
Kennedy/Jenks Consultants
Komline Sanderson Engineering Corporation
Lawler, Matusky & Skelly Engineers, LLP
Limno-Tech Inc.
Lombardo Associates Inc.
Malcolm Pirnie Inc.
McKim & Creed
MEC Analytical Systems Inc.
Metcalf & Eddy Inc.
MWH
New England Organics
Odor & Corrosion Technology Consultants Inc. (OCTC)
Oswald Green, LLC
PA Government Services Inc.
Parametrix Inc.
Parsons
Post, Buckley, Schuh & Jernigan
R&D Engineering/Conestoga Rover & Associates
The RETEC Group
RMC, Inc. (Raines, Melton & Carella)
R.M. Towill Corporation
Ross & Associates Ltd.
Royce Technologies
SAIC Maritime Technical Group
Short Elliott Hendrickson, Inc.
Stantec Consulting Group, Inc.
Stormwater Management, Inc.
Synagro Technologies, Inc.
Tetra Tech Inc.
Trojan Technologies Inc.
URS Corporation
Veolia Water NATC
Wade-Trim Inc.
Weston Solutions Inc.
Woodard & Curran
WRc/D&B, LLC
WWETCO, LLC
Zoeller Pump Company

INDUSTRY

American Electric Power
ChevronTexaco Energy Research & Technology Company
The Coca-Cola Company
Dow Chemical Company
DuPont Company
Eastman Kodak Company
Eli Lilly & Company
Merck & Company Inc.
ONDEO Services
Procter & Gamble Company
PSEG Services Corp.
RWE Thames Water Plc
Severn Trent Services Inc.
United Water Services LLC

Note: List as of 7/1/04

WERF Product Order Form

As a benefit of joining the Water Environment Research Foundation, subscribers are entitled to receive one complimentary copy of all final reports and other products.Additional copies are available at cost (usually $10).To order your complimentary copy of a report, please write "free" in the unit price column.WERF keeps track of all orders.If the charge differs from what is shown here, we will call to confirm the total before processing.

Name ______________________ Title ______________________

Organization ______________________

Address ______________________

City ________ State ________ Zip Code ________ Country ________

Phone ________ Fax ________ Email ________

Stock #	Product	Quantity	Unit Price	Total
			Postage & Handling	
			VA Residents Add 4.5% Sales Tax	
			Canadian Residents Add 7% GST	
			TOTAL	

Method of Payment:

(All orders must be prepaid.)

❑ Check or Money Order Enclosed

❑ Visa ❑ Mastercard ❑ American Express

Account No. ______________________ Exp. Date ________

Signature ______________________

Shipping & Handling:

Amount of Order	United States	Canada & Mexico	All Others
Up to but not more than:	Add:	Add:	Add:
$20.00	$5.00*	$8.00	50% of amount
30.00	5.50	8.00	40% of amount
40.00	6.00	8.00	
50.00	6.50	14.00	
60.00	7.00	14.00	
80.00	8.00	14.00	
100.00	10.00	21.00	
150.00	12.50	28.00	
200.00	15.00	35.00	
More than $200.00	Add 20% of order	Add 20% of order	

*minimum amount for all orders

Note: Please make checks payable to the Water Environment Research Foundation.

To Order (Subscribers Only):

Log on to www.werf.org and click on the "Product Catalog."

Phone: (703) 684-2470
Fax: (703) 299-0742.

WERF
Attn: Subscriber Services
635 Slaters Lane
Alexandria, VA 22314-1177

To Order (Non-Subscribers):

Non-subscribers may be able to order WERF publications either through WEF (www.wef.org) or IWAP (www.iwapublishing.com).Visit WERF's website at www.werf.org for details.

IWA Publishing's authorised EU representative for General Product Safety Regulations is Diane D'Arras, 15 rue Duret, 75116 Paris, France, e-mail: safety@iwap.co.uk.

Printed and bound by CPI Group (UK) Ltd, Croydon, CR0 4YY
12/05/2026
02108892-0009